Intelligence

Nature, Determinants, and Consequences

EDUCATIONAL PSYCHOLOGY

Allen J. Edwards, Series Editor
Department of Psychology
Southwest Missouri State University
Springfield, Missouri

In preparation:

Thomas E. Jordan. Development in the Preschool Years: Birth to Age Five

James R. Layton. The Psychology of Learning to Read

Dale G. Range and James R. Layton (eds.). Early Childhood Education: Theory to Research to Practice

Merlin C. Wittrock (eds.). The Brain and Psychology

Jean Stockard and Patricia A. Schmuck. Sex Equity in Education

Published

Gary D. Phye and Daniel J. Reschly (eds.). School Psychology: Perspectives and Issues

Norman Steinaker and M. Robert Bell. The Experiential Taxonomy: A New Approach to Teaching and Learning

J. P. Das, John R. Kirby, and Ronald F. Jarman. Simultaneous and Successive Cognitive Processes

Herbert J. Klausmeier and Patricia S. Allen. Cognitive Development of Children and Youth: A Longitudinal Study

Victor M. Agruso, Jr. Learning in the Later Years: Principles of Educational Gerontology

Thomas R. Kratochwill (ed.). Single Subject Research: Strategies for Evaluating Change

Kay Pomerance Torshen. The Mastery Approach to Competency-Based Education

Harvey Lesser. Television and the Preschool Child: A Psychological Theory of Instruction and Curriculum Development

Donald J. Treffinger, J. Kent Davis, and Richard E. Ripple (eds.). Handbook on Teaching Educational Psychology

Harry L. Hom, Jr. and Paul A. Robinson (eds.). Psychological Processes in Early Education

J. Nina Lieberman. Playfulness: Its Relationship to Imagination and Creativity

Samuel Ball (ed.). Motivation in Education

Erness Bright Brody and Nathan Brody. Intelligence: Nature, Determinants, and Consequences

The list of titles in this series continues on the last page of this volume

Intelligence

Nature, Determinants, and Consequences

Erness Bright Brody

Department of Science and Humanities
Graduate School of Education
Rutgers—The State University of New Jersey
New Brunswick, New Jersey

Nathan Brody

Department of Psychology
Wesleyan University
Middletown, Connecticut

ACADEMIC PRESS New York San Francisco London 1976

A Subsidiary of Harcourt Brace Jovanovich, Publishers

ACADEMIC PRESS, INC.
111 Fifth Avenue, New York, New York 10003

United Kingdom Edition published by
ACADEMIC PRESS, INC. (LONDON) LTD.
24/28 Oval Road, London NW1

Library of Congress Cataloging in Publication Data

Brody, Erness Bright.
Intelligence.

Bibliography: p.
1. Intellect. 2. Intelligence tests. 3. Nature and nurture. I. Brody, Nathan, joint author. II. Title.
BF431.B6844 153.9 76-39781
ISBN 0-12-134250-6

PRINTED IN THE UNITED STATES OF AMERICA

79 80 81 82 9 8 7 6 5 4 3

For our children Alan and Jennifer

Contents

4

Intelligence and Achievement

5

Determinants of Scores on Tests of Intelligence

6

Group Differences in Intelligence Test Scores

7

Conclusion: The Use of Intelligence Tests

Preface

Recently, as a result of the controversies surrounding group and racial differences in intelligence as well as the social and political implications of these differences, the study of intelligence has become a topic of concern to educators and psychologists.

The purpose of this book is to provide an overview of the field of intelligence to present a balanced account of the major issues.

Chapter 1 serves as the introduction in which the historical development of tests of intelligence and the pioneering work of Spearman and Thurstone are discussed. In Chapter 2 the results of some 70 years of statistically oriented research on the structure of intelligence are considered. The research as presented tends to justify the use of a single numerical index for intelligence test scores. Specific attention is given in Chapter 3 to the quantitative properties of the numerical index and its stability and change over time. Chapters 4 through 6 further develop the meaning of the numerical index by examining its antecedents and consequents. Specifically, Chapter 4 deals with variables, such as education and occupational status, that are assumed to be influenced by intelligence. Chapter 5 addresses the controversial issues surrounding the biological and social determinants of intelligence. The relationships between intelligence test scores and race, family size, and birth order are discussed in Chapter 6. Finally, in Chapter 7 we discuss the uses of intelligence tests and conclude with a skeptical analysis of the utility of such tests.

This book is written primarily for educators, psychologists, students of both disciplines, and the professionals who administer or interpret the results of intelligence tests.

We hope that the material contained in these chapters will serve as background information for future discussions on the topic of intelligence.

The senior author wishes to acknowledge the award of a year's leave granted by the Faculty Academic Study Program and Research Council of Rutgers University which permitted her to work on this manuscript.

1 Antecedents

By 1905 the three researchers who were to have the most profound impact on subsequent work on intelligence had developed their most important concepts. Galton had developed the statistical foundations for the study of individual differences in the last three decades of the nineteenth century and he had initiated a large-scale testing program in 1882. Binet, with Simon, had developed the first useful test of intelligence in 1905, and Spearman had presented, in outline, the theory of general intelligence in 1904.

Not only had these seminal contributions occurred by 1905 but many of the fundamental contemporary questions about intelligence had been raised. These include the question of racial and social class differences in intelligence, the relative contribution of heredity and environment, the potential for change in intelligence either through eugenic or educational procedures, the relationship between intelligence and academic success, the relationship between intelligence as measured by mental tests and physical or external manifestations of intelligence, and perhaps most fundamental of all, the question of whether intelligence is one or many different things. All of these topics are currently the subject of research and controversy.

Sir Frances Galton

Galton was responsible for one of the largest and first attempts to measure individual differences in ability. Galton's work was influenced by Darwin's theory of evolution. The doctrine of the "survival of the fittest" suggested that there are inherited individual differences among members of a species with profound importance for survival. Galton believed (1869), in addition, that there were inherited differences in ability among races and families. He attempted to develop measures of individual differences in ability. Based on observations that individuals of extremely low ability did not excel in distinguishing between different physical stimuli, he assumed (1883) that sensory discriminative capacity would provide an index of general intellectual ability. The first large-scale attempt to measure such differences was begun in 1882. He established an anthropometric laboratory in the South Kensington Museum in London. In addition to the measurement of such characteristics as height and weight, Galton included tests of strength, sensory acuity—e.g., the upper limit of audible sound, and discrimination ability measured by the use of a set of blocks of identical appearance but varying in weight. In all, Galton obtained data with respect to 17 variables for 9337 individuals in conjunction with an international health exhibit.

Galton's contributions to the measurement of individual differences in ability were not exclusively, or for that matter principally, confined to the initiation of the first large-scale testing program. He also made a number of contributions to statistical theory and research methodology that provided the basis for much subsequent research on intelligence as well as other individual difference dimensions. Galton was enormously impressed with the work of the Belgian statistician Quetelet that dealt with the normal curve. Galton pointed out that many individual difference characteristics seemed to be the result of the combination of many small effects which when combined give rise to a normal curve. He pointed out that an individual's score could be defined in terms of his deviation from the central tendency of such a distribution. Galton developed the basic logic of correlation as a measure of relationship between two variables. Finally, he developed both the logic of the twin method (i.e., the comparison of fraternal and identical twins) as a basis for studying genetic versus environmental influences, and the method of comparing foster children with biological children for the same purpose. Galton's work represents the beginning of the English statistical tradition which has dominated theoretical research on intelligence by providing the methodological sophistication required to deal with many theoretical issues. This tradition continues, as we shall see through the work of Spearman, Cyril Burt, and Vernon.

Galton's work also had a profound influence on an American

student of Wilhelm Wundt, James McK. Cattell. Cattell was the first to use the term "mental test" (1890). He used it to describe a battery of tests in use at his laboratory at the University of Pennsylvania. The tests required from 40 to 60 minutes to administer and included measures of sensory acuity for vision and audition, reaction time, sensitivity to pain, color preferences, memory, and imagery. The tests dealt with sensory functions and relatively simple reactions.

Alfred Binet

In 1896 Binet and Henri, prompted by their disagreement with the measurement of ability as represented by Galton and Cattell, wrote a paper to express their views.

Binet and Henry (1896) had two principal criticisms of the type of mental testing developed by Cattell. First, they believed that the tests were weighted too heavily in the direction of sensory functioning and simple psychological processes. Second, they argued that the tests failed to contain a sufficiently varied sample of measures pertaining to diverse mental faculties. In this paper they presented an outline of a mental test that they believed would be more adequate. It would sample a variety of psychological functions in a single relatively brief session and it would emphasize the superior or higher mental abilities. They included tests of such abilities as memory for a variety of materials (e.g., musical notes, digits, and words), tests of imagery, tests of imagination, attention, comprehension, suggestibility, aesthetic appreciation, moral sentiments, muscular force, motor skills, and judgment of visual space. In all, 10 faculties were to be measured.

The 1905 test was finally developed in response to the work of a commission appointed by the French Minister of Public Instruction (Binet & Simon, 1905a–c). The commission was concerned with the education of mentally defective children who were not able to profit from instruction in regular public school classes. The commission called for the development of objective procedures for the discovery and selection of such children. Binet and Simon sought to develop an objective test (i.e., a test whose administration and scoring would be standardized) that would be able to meet the needs as outlined by the commission. The test was the first one to attempt to develop a metrical (quantitative) concept of intelligence. The Binet–Simon test contained over 30 subtests, which were easily administered and scored.

The final tests were compatible with the programmatic concepts of the 1896 Binet and Henri paper. They included a great diversity of content and there was an emphasis on the "higher" mental processes. In one respect, however, the test was different from that called for in 1896. The higher mental processes that were to be included in 1896 were suggested principally by the theoretical conceptions of faculty psychology. The 1905 test reflected the outcome of a decade of empirical research and included, for the most part, items for which the probability of successful performance increased with chronological age. The tests included the naming of objects, digit span memory, word definitions, and comprehension items. With minor modifications, some of the items on the 1905 examination are included in the latest edition of the widely used Stanford–Binet test of intelligence. These include such items as identifying parts of the body, digit span, memory for sentences, and finding the similarity between two things. The 1905 tests were arranged in order of difficulty and a crude scoring system was used based on the highest item attained. Various types of mental defectives could not answer questions beyond a certain level. The concept of mental age was not specifically presented.

Binet and Simon recognized that ability to perform correctly increased with age and that the performance of the mentally defective resembled that of the "normal" child of a younger age. The test, though crude, is virtually identical to contemporary tests in such essential features as the emphasis on complex mental functioning assessed by brief measures which are administered and scored by standardized procedures, and the use of items whose difficulty level is age-related.

The 1905 scale was followed by a revision (Binet & Simon, 1908). The 1908 tests were designed to be used to distinguish among the intellectual abilities of normal children as well as to distinguish between the intellectual abilities of normal and mentally defective children. Also, the concept of mental age was made more explicit. Tests were grouped according to age level, and the mental age of a child was defined as the highest age level at which the child was able to pass all items but one plus a credit of 1 year for every five items passed at a more advanced age. Table 1.1 presents an outline of the test.

In 1911 Binet published a revision of the 1908 scale. The revision followed the essential outline of the 1908 scale introducing modifications based on the experience of various examiners with the earlier scale. The 1911 version had five items at each age (except for age 4, which, probably by oversight, was left with the same four tests). The difficulty level of a number of items was changed and the rules for determining mental age were also changed. Although Binet's tests of 1908 and 1911 provided explicit procedures for the determination of mental age, they did not include a procedure for the determination of an intelligence quotient (IQ), which permits comparisons of intellectual ability of individuals at different age levels.

Binet made contributions to two other research topics in the area of intelligence. He dealt with the attempt to raise intellectual performance by environmental intervention. He developed a series of procedures referred to as "mental orthopedics": These procedures were designed to increase facility on certain component skills with the expectation that such training would improve overall intellectual functions. Binet's beliefs about the efficacy of training can be contrasted with Galton's views. Galton believed that the most efficacious method of improving mental ability was by eugenic procedures. The contrast between eugenic intervention and environmental interventions continues, as we shall see, through the present time.

Binet's mental orthopedics was the forerunner of many contemporary efforts to increase intelligence test scores by specific environmental interventions.

Binet was also concerned with the attempt to relate mental ability to physical characteristics. He did extensive studies relating head measurements and physiognomy to mental ability. These studies may be conceived as the forerunners of contemporary efforts to relate scores on intelligence tests to physiological indices.

Binet never developed a final theoretical conception of intelligence. His original notions about the nature of intelligence were dominated by the concepts of faculty psychology, suggesting that intelligence was defined by a diverse set of independent abilities. Yet he eventually developed a test of intelligence that provided for the computation of a single index, implying that intelligence was one thing. This was an issue Binet never faced squarely. It was Spearman (1904b) who proposed that intelligence be considered one general thing and that tests could be constructed to provide for a single pooled index across items. Yet Binet (1905) was somewhat skeptical of Spearman's early work. Tuddenham (1962) aptly summarizes Binet's views and distinguishes them from the theory that intelligence is one single thing: "Regarding intelligence as a product of many abilities, Binet sought in his tests to measure not an entity or single dimension—'general intelligence'—but rather an average level—'intelligence in general' [p. 489]."

TABLE 1.1
Items from the 1908 Binet Scale

Age 3 years

1. Points to nose, eyes, mouth.
2. Repeats sentences of six syllables.
3. Repeats two digits.
4. Enumerates objects in a picture.
5. Gives family name.

Age 4 years

1. Knows sex.
2. Names certain familiar objects shown to him: key, pocketknife, and a penny.
3. Repeats three digits.
4. Indicates which is the longer of two lines 5 and 6 cm in length.

Age 5 years

1. Indicates the heavier of two cubes (of 3 and 12 gm and also of 6 and 15 gm).
2. Copies a square, using pen and ink.
3. Constructs a rectangle from two pieces of cardboard, having a model to look at.
4. Counts four pennies.

Age 6 years

1. Knows right and left as shown by indicating right hand and left ear.
2. Repeats sentence of 16 syllables.
3. Chooses the prettier in each of three pairs of faces.
4. Defines familiar objects in terms of use.
5. Executes a triple order.
6. Knows age.
7. Knows morning and afternoon.

Age 7 years

1. Tells what is missing in unfinished pictures.
2. Knows number of fingers on each hand and on both hands without counting them.
3. Copies "The little Paul" with pen and ink.
4. Copies a diamond, using pen and ink.
5. Repeats five digits.
6. Describes pictures as scenes.
7. Counts 13 pennies.
8. Knows names of four common coins.

Age 8 years

1. Reads a passage and remembers two items.
2. Counts up the value of three simple and two double sous (or 1¢ and 2¢ stamps in American scales).
3. Names four colors—red, yellow, blue, green.
4. Counts backward from 20 to 0.
5. Writes short sentence from dictation, using pen and ink.
6. Gives differences between two objects from memory.

Continued

TABLE 1.1—*Continued*

Age 9 years

1. Knows date—day of week and of month, also month and year.
2. Recites days of week.
3. Makes change on 4¢ out of 20¢ in simple play-store transactions.
4. Gives definitions superior to use (familiar objects).
5. Reads a passage and remembers six items.
6. Arranges five blocks in order of weight.

Age 10 years

1. Names in order the months of the year.
2. Recognizes all the (nine) pieces of money.
3. Constructs a sentence to include three given words—Paris, fortune, gutter. Two unified sentences are acceptable (passed by only about 50%).
4. Answers easy comprehension questions.
5. Answers hard comprehension questions. Only about half of the 10-year-olds get the majority of these correct.

Age 11 years

1. Points out absurdities in contradictory statements.
2. Sentence construction as in 3 for age 10 years. Hardly one-fourth pass the test at 10 years, whereas all do at 11 years of age.
3. Names 60 words in 3 minutes.
4. Defines abstract terms—charity, justice, kindness.
5. Puts words, arranged in a random order, into a sentence.

Age 12 years

1. Repeats seven digits.
2. Finds in 1 minute three rhymes for a given word—obedience.
3. Repeats a sentence of 26 syllables.
4. Answers problem questions—a common-sense test.
5. Gives interpretation of pictures.

Age 13 years

1. Draws the design that would be made by cutting a triangular piece from the once-folded edge of a quarto-folded paper.
2. Rearranges in imagination the relationship of two triangles and draws the results as they would appear.
3. Gives differences between pairs of abstract terms, as pride and pretension.

Charles Spearman

The Background of Spearman's Papers of 1904

We have already seen that Binet and Henri's paper of 1896 was prompted by their disagreement with the widespread testing programs similar to that undertaken by Galton. Spearman's papers of 1904 were prompted by his disagreement with American disagreements with the mental test procedures of the 1890s. Two well-known investigations by Sharp (1898,1899) and Wissler (1901) had done much to turn American academic psychology away from the method of mental tests. The mental test movement developed in large measure outside the auspices of American academic departments—perhaps to the detriment of both. Sharp's study, done in Tichener's laboratory at Cornell University, was an attempt to evaluate the efficacy of the type of tests advocated by Binet and Henri. Her conclusions about the tests, though carefully balanced, were negative. She believed that the tests required refinement and did not meet the meticulous standards of rigor in vogue in experimental laboratories. Also, she thought that little could be learned from the tests that would be of value to investigators using Titchener's introspective methods. And, most critical of all, her perusal of the data suggested that there was a lack of correspondence in the individual difference scores obtained on the various tests. If the tests were substantially unrelated, then there was no basis for combining their results into a single common index.

Sharp's results were buttressed by Wissler's analysis of data collected in Cattell's laboratory. Wissler made use of Pearson's product-moment correlation as a means of evaluating the relationship among tests. He found that correlations among mental tests ranged from −.28 to +.39 with an average correlation of .09, and their average correlation with intelligence as measured by class grades was .06. Wissler, as a result, concluded that the various measures were substantially unrelated and did not provide the basis for a common pooled index.

With hindsight, we realize that the Sharp and Wissler studies, though enormously influential, were extremely inadequate. Sharp used only seven subjects—each of whom was a graduate student. Since all of the subjects were extremely high in ability, we would not expect to find much variation in test scores among them. The "restriction in range of talent" would, of necessity, reduce the correlations among the tests. The same criticism can be applied to Wissler's study, which dealt only with the scores of college students. Also, Wissler's study dealt with measurements of simple sensory tasks rather than the tasks advocated by Binet and Henri.

Spearman (1904a) had two additional criticisms of Wissler's study. First, he believed that the tests were not administered in a proper or rigorous fashion. He noted that subjects were tested three at a time rather than individually and that 22 different tests were completed in 45 minutes. Second, Spearman pointed out that Wissler and earlier investigators failed to consider the phenomenon of "attenuation." Spearman pointed out that the correlation between two measures was less than the hypothetical true correlation between them due to the presence of errors which served to depress or attenuate the observed correlation. The formula presented by Spearman to represent the influence of attenuation would be accepted by contemporary theorists—although the basis for computing the quantities contained in the correlation would not be the same. Spearman's formula was:

$$\hat{r}_{xy} = r_{xy}/(r_{xx'}r_{yy'})^{1/2},$$

where $\hat{r}_{xy}$ (in contemporary terms) would be understood as the estimated true relationship between two measures—x and y—r_{xy} is the obtained correlation between x and y, and $r_{xx'}$ and $r_{yy'}$ are the respective reliabilities of the tests.

The formula implies that observed correlations are always lower than "true correlations" and that the difference between observed and true correlations is a function of the unreliability (or extent of errors of measurement) of each of the tests. The brief, casually administered tests in Cattell's battery were quite probably of very low reliability. Hence, the true correlations among them and between them and grades (which are themselves not perfectly reliable) were probably substantially greater than those obtained by Wissler.

Spearman (1904b) reported the results of a study relating school achievement to sensory discrimination ability in which he came to the somewhat astounding conclusion that when corrected for attenuation, the appropriate relationship between these measures was a correlation of 1.00. Although this relationship was wisely, and probably correctly, ignored in Spearman's subsequent work it was in point of fact the historical foundation of his general theory of intelligence.

The Two-Factor Theory of Intelligence

Spearman suggested that each measure or test in the intelligence domain was based on two factors—*g* and *s*. The factor *g* represented that which the test had in common with all other tests of ability. The factor *s* represented a separate factor that was initially assumed to be specific to each test (or, perhaps, to be present also in virtually identical tests). If the *s* factors were truly specific and present in one and only one test, then the basis for the relationship between two tests would be the amount of *g* they shared in common. Tests, however, may differ in the extent to which they are measures of *g*.

Spearman's theory has the great advantages of being explicit and of suggesting a number of empirical consequences that can be tested. If the theory is correct, all correlations between measures of intelligence (excluding considerations of sampling error among correlations) should be positive. Also, it should always be possible to arrange the tests in a hierarchy such as that presented in Table 1.2.

Note that in Table 1.2 the correlations decrease going down a column or across a row. This comes about because the test in the upper left corner of the matrix, No. 1, is that test with the highest amount of *g*. The next test, No. 2, represents the test with the next highest amount of *g*. The next test, No. 3, has the next highest amount of *g*. The correlation between 1 and 2 will be higher than any other correlation in the table, since these two tests have the highest amount of *g* in common. Also, the correlation between 1 and 3 will be higher than the correlation between 2 and 3 on the general assumption that the correlation is a function of the degree of *g* shared in common by the tests.

The correlation between any two tests is given by the formula $r_{12} = r_{1g} \cdot r_{2g}$ where r_{1g} and r_{2g} represent the correlations between tests 1 and 2 and *g*, respectively. Therefore if the "*g* loading" of two tests is known, the expected correlation between them can be predicted.

Subsequently, Spearman was to point out that if the two-factor theory of intelligence was correct, the "law of tetrad differences" would hold for correlations among any four measures of intellectual ability (see Spearman, 1927, for a comprehensive discussion of his early research). The law of tetrad differences may be stated as follows:

$$r_{12}r_{34} - r_{14}r_{23} = 0$$

TABLE 1.2
A Hierarchical Arrangement of Correlations Derived from Spearman's Theory

Tests	1	2	3	4	5
1		.72	.63	.54	.45
2			.56	.48	.40
3				.42	.35
4					.30
5					

Spearman's theory not only could be used to generate precise, testable consequences about the expected relationship that would obtain among correlations of ability measures, it also provided a basis for the selection of optimal measures of ability. Rather than the suggestion that test constructors should rely on intuition or a priori judgment in the selection of tests, Spearman's theory implies that those measures for which the ratio of g/s is a maximum should be selected. Because on this theory the test constructor wishes to obtain a measure of g, he will be able to obtain his most accurate estimate if he restricts his tests to those that measure g rather than specific ability.

Emendations on Spearman's Theory and the Problem of "Group Factors"

Spearman refused to identify g with "intelligence" inasmuch as he considered the latter term vague. He did however assert that various tests of intelligence that were in use in the first two decades of the twentieth century were in fact measures of g. The least inferential conception of g to which Spearman adhered was that g was simply that factor common to all measures of intellectual abilities. However, he went on to postulate both a more inferential psychological characterization of g and a still more hypothetical conception of g as mental energy. By an examination of the tests that seemed to most clearly measure g, Spearman came to the view that g was principally related to the ability to perform intellectual operations he called "eduction of relations and correlates" which he defined as follows:

> The eduction of relations . . . when a person has in mind any two or more ideas (using this word to embrace any items of mental content . . .) he has more or less power to bring to mind any relations that essentially hold between them.
>
> It is instanced whenever a person becomes aware, say, that beer tastes something like weak quinine . . . or that the proposition "All A is B" proves the proposition "Some A is B". . . .
>
> The eduction of correlates . . . when a person has in mind any idea together with a relation, he has more or less power to bring up into mind the correlative idea.
>
> For example, let anyone hear a musical note and try to imagine the note a fifth higher. . . . Or let him notice the relation of horizontal to vertical length in Fig. 1 below, and then try to draw in Fig. 2. . . a horizontal length in the same relation to the vertical one as before. . .

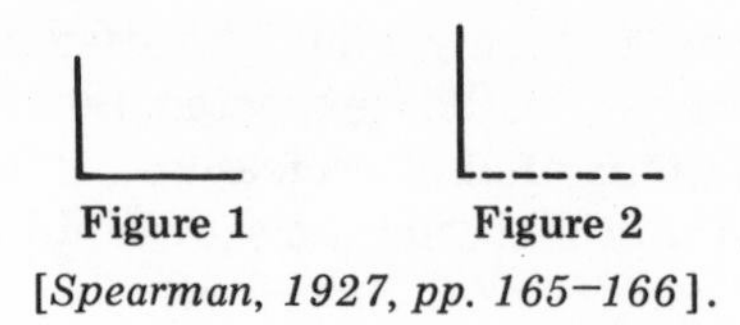

Figure 1 Figure 2

[*Spearman, 1927, pp. 165–166*].

The two types of eduction just defined combined with the "law" that asserts that a person has more or less power to observe what goes on in his own mind provides three qualitative laws of cognition: These three laws, in turn, give rise to a number of other subdivisions and questions such as "(*a*) the different classes of relation that are cognizable (*b*) the different kinds of fundaments that can enter into these relations and (*c*) the varying kinds and degrees of complexity in which such relations and fundaments can be conjoined [Spearman, 1927, p. 411]."

It is apparent that Spearman's theoretical conception of intelligence is quite complicated and extends beyond the two-factor theory developed in 1904. There is little relation between the specialized algebra suggested by the two-factor theory and the more complex theoretical model of intelligence that Spearman presented in the 1920s (Spearman, 1927).[1]

The most difficult empirical issue faced by Spearman's theory, and the point at which it is generally assumed to have been at variance with empirical results, stemmed from the existence of "group factors." Spearman recognized that the law of tetrad differences implied by his theory was not completely in accord with empirical results. Of course it would be natural to expect some degree of discrepancy between theory and results due to sampling error. That is, since obtained correlations based on a sample of subjects only approximate the hypothetical correlations between tests in the population from which the sample was derived, it is to be expected that there would be sampling errors which would tend to produce deviations from the theoretically expected value of 0 in all tetrad differences. However, sampling error may be estimated and the law of tetrad differences can be modified to accommodate their influences. A second more serious problem occurs when two measures or tests are so similar that their similarity (or correlation)

[1] Spearman's theory is further complicated by the postulation of a number of other cognitive abilities that are wholly or partially independent of *g*. These include such characteristics as retentivity (memory) and perseveration.

includes a common *s* (shared variance) as well as their similarity in *g*. Such tests would tend to correlate higher than would be expected on the basis of their relationship to *g*. An example given by Spearman is two tests of cancellation—one involving the cancellation of all *a*'s on a page, the other the cancellation of all *e*'s. In this case the tests are trivially similar. However, if we allow that different tests that are similar and almost identical may share *s* we now have rendered the two-factor theory untestable. That is, unless there is an independent way of defining similarity, we can always account for excessively positive correlations leading to the disconfirmation of the law of tetrad differences by appealing to the notion of similarity between tests. This difficulty is far more profound than the trivial example cited above might indicate. As early as 1906 Krueger and Spearman had noted that certain relatively dissimilar tests could have correlations higher than the values expected on the basis of their *g* loadings. They raised the possibility that a group of diverse measures might relate together to form a unitary ability—in their case they speculated about the existence of a more or less unitary memorization ability. Such overlapping patterns of correlations give rise to "group factors" (as distinguished from the general factor or *g*) and are defined as factors that relate to more than one of a set of ability measures but, unlike *g*, are not present in all measures of ability. The existence of group factors, of course, is incompatible with Spearman's two-factor theory. Spearman, and psychologists working in England who have been influenced by him, have, for the most part, tended to recognize the existence of such factors but have also tended to deemphasize their importance. On a purely theoretical level, there is no reason why one cannot incorporate group factors into an expanded version of Spearman's two-factor theory. Such a theory would assert that scores on a test are determined by its loading on *g*, its loading on one (or possibly more than one special ability represented by a group factor) and *s*. The correlation between a pair of tests would then be due to their common *g* and their shared variance on any group factor. However, this apparently innocuous theoretical emendation does not fit simply or elegantly into Spearman's algebra. The virtue of Spearman's original theory is that it provides for an elegant integration of theory and methodology permitting one to test the theory rigorously be examining obtained patterns of correlations. No such test, and indeed no general procedure for the precise description and measurement of group factors, can be obtained with the use of Spearman's statistical procedures.

Thurstone and the Doctrine of Simple Structure

Thurstone (1931), working in America, is generally credited with developing statistical procedures capable of rigorously dealing with group factors.[2] Spearman's methods of analysis were predicated on the assumption that only one factor was present in a matrix of correlations between all possible pairs of tests. In contrast, Thurstone's statistical procedures permitted one to discover empirically the number of factors present in the matrix. In addition, Thurstone also provided procedures that would enable him to define the factors. The definition of a factor is given in terms of the subset of tests that load on it. Thurstone's statistical methodology, like Spearman's before him, was an outgrowth of a set of theoretical assumptions about the nature of abilities. Thurstone began with the assumption that performance on a test of ability was dependent upon a certain number of fundamental or primary abilities (his term for the psychological characteristic represented by a group factor). Second, he assumed that the number of primary abilities (or group factors that must be postulated) will be less than the number of tests administered (assuming, of course, that a relatively heterogeneous set of tests is included). This principle is an expression of a rule of parsimony according to which there would be little gain in explanatory power or economy in postulating that the number of abilities is equal to the number of tests in a set. Psychologically, this assumption is equivalent to the notion that there exist abilities that are not specific to a particular test but that enter into performance on a variety of tests. Third, Thurstone assumes that performance on a particular test does not involve all of the primary abilities that must be postulated to explain performance on all of the tests. These assumptions provide a criterion, called simple structure, that guides the solution of a mathematical problem. Given a matrix of correlations composed of correlations among all possible pairs of tests, the factor analyst seeks to discover the minimal number of factors that must be postulated to explain performance on the tests. Once the factors are chosen we can define a new matrix, as in Table 1.3, consisting of the loadings (hypothetical correlations between the tests and the factors) of the tests on the factors.

[2] There were, however, a number of precursors of Thurstone's statistical discoveries. Vernon (1961) suggests that Burt (1917) had already discovered the basic methodology used by Thurstone. Similarly, Cattell (1971, p. 25) indicates that J. C. M. Garnett had presented the idea in the Proceedings of the Royal Society in 1919.

TABLE 1.3
A Hypothetical Factor Structure
Indicating Loadings of Tests on Factors

	Factors		
Tests	A	B	C
1	.73	.21	.10
2	.65	.18	.10
3	.61	.27	.25
4	.10	.53	.11
5	.13	.61	.02
6	.00	.79	.01
7	.12	.02	.44
8	.05	.05	.39
9	.03	.10	.55

The principle of simple structure seeks to maximize the number of zero loadings in such a table. There should be one or more zero entries in each row. Where the loading of a test on a factor is zero, this implies that the ability represented by that factor is not involved in performance on the test. Also, the factors should be defined such that the loadings of the tests on some of the factors are maximized. The ideal toward which the factor analyst strives is a matrix in which each test has high loadings on a small number of factors (one, in the limiting case where test performance is determined by a single ability) and zero loadings on all other factors.

Although there is no mathematically unique solution to the factor analysis of a matrix of correlations, the notion of simple structure provides a guideline or criterion toward which the analysis is directed.

Thurstone developed this type of factor analysis in 1931. His first large-scale study of abilities using this method was published in 1938. The study consisted of the administration of a battery of 56 tests to 218 college students. Nine factors were tentatively identified. These were *S*-Spatial, *P*-Perceptual, *N*-Numerical, *V*-Verbal Relations, *W*-Words, *M*-Memory, *I*-Induction, *R*-Arithmetic Reasoning, and *D*-Deduction. Thurstone thought that the two last factors were less clearly defined than the others. The definition of a factor mathematically is given by the loadings of the tests on the factor. Psychologically, the factor is defined in terms of an attempt to comprehend the basis for the underlying unity in the tests that load most substantially on the factor. The process of discovering the underlying unity shared by diverse tests may vary in difficulty. Thurstone found the two last factors, *R* and *D*, derived in his study less well defined principally because he was unsure about the underlying unity present in the diverse tests whose loadings jointly defined the factor. The interpretation of a factor may be conceived of as the suggestion of a hypothesis that is subject to empirical confirmation. The interpretation ought to suggest the kinds of tests that would load on the factor in future investigations. Suppose that it is assumed that three tests load positively on a factor tentatively interpreted as *D*-Deduction. Then in a new investigation in which these three tests are included plus a new test that is assumed to measure deductive ability, a factor should be derived defined by positive loadings of the three previous tests plus the new test of deductive ability.

Spearman and Thurstone

What is remarkable about Thurstone's investigation is that it carries us back to the issue presumably disposed of by Spearman in 1904, namely, is intelligence one or many things and can scores on intelligence tests properly be defined in terms of a single number? In dealing with this issue, which is still the subject of debate, one is struck by the extent to which both Spearman and Thurstone developed methodologies that were suited to their theoretical assumptions. The method of tetrad differences is valid only where there is a single factor in the matrix. The principle of simple structure is useful only if the correlations in the matrix are to be accounted for by a set of group factors. If *g* exists, then simple structure cannot exist since there will be at least one factor (representing *g*) where every test has nonzero loadings.

At first, Thurstone was inclined to argue that his results were in complete contradiction to Spearman's. Furthermore, he believed that his results contradicted the use of a single IQ index to describe an individual's ability. Rather, an individual's ability ought to be described, according to Thurstone, in terms of a profile representing his scores on the primary mental abilities.

Actually, the disagreement between Thurstone and Spearman is not as great as it would appear on the basis of this discussion. In order to explain this, it is necessary to distinguish between orthogonal and oblique rotations and first- and second-order factors. It is possible to perform a factor analysis resulting in a set of factors that are themselves unrelated or uncorrelated. Such factors are called orthogonal. Oblique factors are those that are themselves correlated.

Some factor analysts believe that only orthogonal factors are legitimate, because only they represent truly independent and unrelated explanatory constructs. Other factor analysts permit oblique factor solutions, because, they argue, there can be logically distinct explanatory constructs that are nevertheless quantitatively related to each other. An example cited by Thurstone would be height and weight—obviously related and yet obviously conceptually distinct. Thurstone found it necessary to use oblique factors in order to obtain solutions that fit the criterion of simple structure. If there is a set of factors that are themselves correlated, then a correlation matrix representing the correlations among all possible pairs of factors can be formed. And this correlation matrix can itself be factor analyzed. This is called a second-order factor analysis.

As early as 1940, Raymond Cattell (Cattell, 1971), a former research assistant of Spearman's who emigrated to America, had indicated that a second-order factor analysis would result in a rapprochement between the views of Spearman and Thurstone. On this view, *g* would emerge as a second-order factor. Thurstone was in substantial agreement with this view. He did, however, point out that while second-order factor analysis did produce a factor similar to Spearman's *g*, it was not the only second factor that emerged.

On the other hand, Spearman could point out that the results of Thurstone's factor-analytic investigation of ability did not really contradict the existence of *g*. For one thing, the correlations among the tests in Thurstone's battery were overwhelmingly positive. The median correlation, uncorrected for attenuation, was +.35. Note further that the sample was one that had a restricted range of talent, thereby depressing the value of the correlations. Furthermore, the general reliance on oblique solutions, it could be argued, was due to the existence of *g*. That is, if orthogonal solutions were demanded it would be difficult to obtain simple structure because tests would tend to have nonzero loadings on more than one factor. In short, it could be argued that Thurstone's studies had not shown that *g* did not exist but rather that *g* could be distributed and divided into components that were themselves positively related. Since Thurstone recognized the existence of *g* and Spearman recognized the existence of group factors representing special abilities, it is apparent that the difference between Spearman and the British school and Thurstone and his followers was largely a matter of emphasis. Spearman continued to assign primary importance to *g* and to consider group factors or primary abilities as of minor importance. For Thurstone, the reverse was true.

2

The Structure of Intellect

In this chapter we shall consider the contemporary theoretical efforts to build a theory that explains the structure of intellect. We shall see that in large measure contemporary theories are outgrowths of the earlier work of Thurstone and Spearman and that their differences in emphasis exist, if anything, in more extreme form today.

Raymond B. Cattell

Cattell's theory of intelligence is a contemporary synthesis of the Spearman and Thurstone traditions. Like Spearman, he is an advocate of the importance of *g*. And, like Thurstone, he believes in the use of oblique rotations and derives *g* as a second-order factor.

Cattell's theory is quite complex and exists at several levels of generality. Parts of the theory are reasonably well grounded in empirical work. Other parts represent reasonably modest extensions and interpretations of current research. Still other parts are, at this stage of research, frankly speculative. Cattell's book (1971), *Abilities: Their structure, growth, and action*, represents both a synthesis of existing research on intelligence and a number of suggestions for future research.

Cattell's analysis of intelligence takes as its starting point an analysis of primary abilities quite analogous to the procedures followed by Thurstone. That is, there is an attempt to sample a wide range of tests and to use oblique rotations in order to satisfy the criterion of simple structure. Table 2.1 presents an outline of a list of primary abilities recognized by Cattell. Note that the list includes many of the abilities that Thurstone reported in his 1938 monograph. The difference between Cattell and Thurstone at this point is one of emphasis and interest. Thurstone was primarily concerned with discovering primary abilities and describing individual differences in ability in terms of an individual's scores on the set of primary ability factors. Cattell has done relatively little research directed toward the discovery and description of primary abilities per se. Tests that load heavily on various primary abilities are used by him as a basis for second-order factor analysis, permitting him to describe individual differences in ability at a more abstract and general level.

Cattell (Horn & Cattell, 1966) has reported the results of a second-order factor analysis of abilities in which five second-order factors were derived. Two of the five factors represent a division of Spearman's g into two components—g_f which stands for fluid ability and represents the basic biological capacity of the individual and g_c which stands for crystallized ability and represents the type of ability

TABLE 2.1
A Tentative List of Empirically Based Primary Ability Concept[a]

1. Verbal ability
2. Numerical ability (basic manipulation facility; not mathematics)
3. Spatial ability
4. Perceptual speed (figural identification)
5. Speed of closure (visual cognition, Gestalt perception)
6. Inductive reasoning (general reasoning)
7. Deductive reasoning (logical evaluation)
8. Rote memory (associative memory)
9. Mechanical knowledge and skill
10. Word fluency
11. Ideational fluency
12. Restructuring closure (flexibility of closure)
13. Flexibility versus firmness (originality)
14. General motor coordination (psychomotor coordination)
15. Manual dexterity
16. Musical pitch and tonal sensitivity
17. Representational drawing skill

[a]Based on Cattell (1971).

measured by most standardized tests of intelligence. It is assumed to represent the effect of acculturation upon intellectual ability. The distinction between g_f and g_c is the most fully developed aspect of Cattell's theory and we shall, subsequently, discuss it at length. The other second-order factors extracted in the Horn and Cattell study were g_v—power of visualization—a factor assumed to reflect the role of visualization ability in solving diverse problems; g_r—retrieval capacity or general fluency, which refers to the ability to retrieve or recall many different items rapidly from mental storage; and g_s—cognitive speed, the ability to perform well in speeded situations which are somewhat less complex than those that are good measures of Spearman's g or g_f for Cattell. The reason for this distinction between g_s and speed in complex performances actually goes back to an observation of Spearman's made in 1904 (Spearman, 1904b). Spearman noted that scores on test performances in complex task under conditions demanding speeded performance correlated quite well with scores from tests where speeded performance was not required.

The delineation of the set of second-order abilities (with the exception of the g_f–g_c distinction) rests only on one published study. The precise pattern of abilities that emerges from a second-order factor analysis requires replication and extension using different batteries of tests. However, what we can consider as being established from the point of view of Cattell's theory is the generalization that second-order factor analysis of first-order ability factors obliquely rotated to simple structure leads to several factors including g_f, g_c and other factors representing general features of cognitive ability.

The g_f–g_c Theory

The g_f–g_c theory was first presented in 1941. It was based in part on observations in Spearman's laboratory made in the 1930s which indicated that tests involving perceptual classifications and analogies were highly intercorrelated and seemed to be particularly clear measures of g. On the other hand, tests that seemed to be more related to scholastic achievement and to reflect knowledge acquired in a school situation tended to load somewhat lower in g and to be somewhat separate from the tests of perceptual classification and analogies. The tests that were relatively pure measures of g were similar to those that came to be used in "culture-fair" or "culture-reduced" tests.

One procedure that can be followed in constructing such a test is to present items involving the ability to combine relatively simple and common elements in complex ways. Example of such items are presented in Table 2.2. Such items are assumed to be culture reduced because the cognitive elements that must be combined are assumed to be available to all members of the culture and, in addition, the task posed to the subject is assumed to be relatively novel for all members of the culture. Such items evidently call for the ability to educe relations and correlates in Spearman's sense. Such items are assumed to measure g_f. This type of item may be contrasted with a vocabulary test, which is not culture reduced or culture fair in that performance on a vocabulary test obviously depends on the cultural experience and schooling of an individual. Performance on a vocabulary test would be more reflective of g_c than g_f.

In order to delineate more precisely the distinction between g_f and g_c we will begin with a description of the factor analyses performed by Cattell and his associates which serve as the foundations for the description of each of the constructs. Horn (1968) has summarized the results of a variety of second-order factor analyses in which the distinction between g_f and g_c has appeared. Table 2.3 presents Horn's summary of these studies. The factor coefficients appearing in Table 2.3 represent average loadings taken from several investigations. An examination of Table 2.3 permits us to obtain a somewhat clearer understanding of the characteristics of g_f and g_c. Note that g_f is defined relatively uniquely by three tests (symbolized as CFR, 17s, and 17). These are tests that have moderately high loadings on g_f and loadings that approximate zero on g_c. Presumably these tests reflect a basic biological capacity to learn, which is relatively uninfluenced by acculturation. Two of the three types of factors that load heavily on g_f in Horn's summary are closely related to Spearman's view of g as the capacity to educe correlates and relations—indicating that g_f is conceptually analogous to Spearman's g. The g_c factor is best and most clearly defined by V or verbal comprehension—a factor that is presumed to depend heavily on acculturation. Note further that a number of first-order factors are about equally loaded on both g_f and g_c. The fact that a number of tests are related to both the g_f and g_c second-order factors is indicative of the fact that these two factors are themselves correlated—the correlation is assumed to be between .4 and .5. One suspects that this degree of separation is maximal. That is, that the correlation occurs even when final factor solutions are chosen that are designed to maximize the distinction or to separate optimally the two factors. The substantial relationship between g_f and g_c may also

TABLE 2.2
Examples of Five Culture-Fair Perceptual, Relation-Education Subtests of Proven Validity for Fluid Intelligence[a]

Choose one to fill dotted square.

Series

   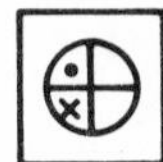     

Choose odd one.

Classification

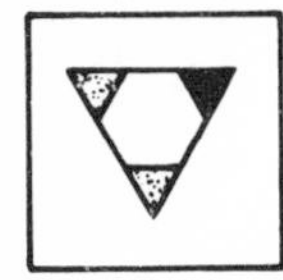 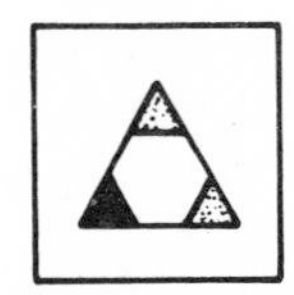   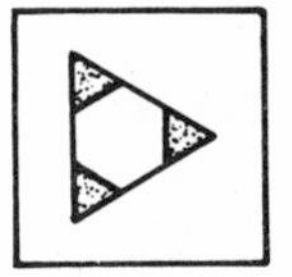

Choose one wherein dot could be placed as in item on left.

Topology

Choose one to complete analogy.

Analogies

 is to as  is to    

Choose one to fill empty square at left.

Matrices

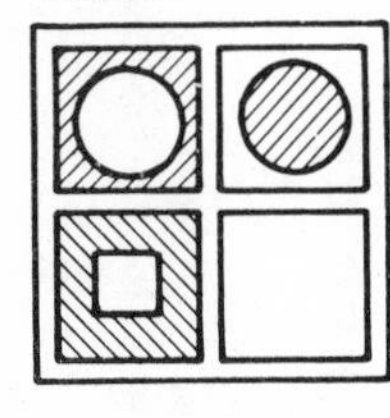 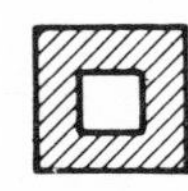  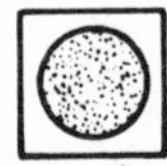   

From Form B, Scales II and III, IPAT Culture-Fair Test. By kind permission of the Institute of Personality and Ability Testing, 1602 Coronado Drive, Champaign, Illinois

Analogies section from Cattell Scale II, Harrap & Co.

---

[a]From Cattell (1971).

TABLE 2.3

Summary of Some Results from Studies in Which g_f and g_c Factors Have Been Identified[a]

Symbol	Behavioral indicant	Approximate factor coefficient		Symbol	Behavioral indicant	Approximate factor coefficient	
		g_f	g_c			g_f	g_c
CFR	*Figural relations.* Education of a relation when this is shown among common figures, as in a matrices test.	.57	.01	Rs	*Formal reasoning.* Arriving at a conclusion in accordance with a formal reasoning process, as in a syllogistic reasoning test.	.31	.41
Ms	*Memory span.* Reproduction of several numbers or letters presented briefly either visually or orally.	.50	.00	N	*Number facility.* Quick and accurate use of arithmetical operations, such as addition, subtraction, and multiplication.	.21	.29
I	*Induction.* Education of a correlate from relations shown in a series of letters, numbers, or figures, as in a letter series test.	.41	.06	EMS	*Experiential evaluation.* Solving problems involving protocol and requiring diplomacy, as in a social relations test.	.08	.43
R	*General reasoning.* Solving problems of area, rate, finance, etc., as in an arithmetic reasoning test.	.31	.34	V	*Verbal comprehension.* Advanced understanding of language, as measured in a vocabulary or reading test.	.08	.68
CMR	*Semantic relations* Education of a relation when this is shown among words, as in an analogies test.	.37	.43				

[a]From Horn, J.L. Organization of abilities and the development of intelligence *Psychological Review*, *75*, 249. Copyright 1968 by The American Psychological Association. Reproduced by permission.

be considered as evidence for Spearman's original position in postulating a single *g*. And we shall see that there is a sense in which Cattell's theory is even closer to Spearman's than we have yet indicated. We can anticipate this point by indicating that in a rough sense g_c is dependent upon g_f. That is, the acquisition of intellectual skills as a result of acculturation is dependent not only on the quality of one's cultural and educational experiences but also on the level of fluid ability an individual has which permits him to benefit from the educational experiences made available to him.

An examination of the average factor loadings reported in Table 2.3 indicates that there is no measure or primary factor that is unique and substantially defined by either g_f or g_c. We can make this point clearer by noting that the percentage of variance accounted for by a correlation of a particular value is given by r^2. Therefore, we can assert that the primary factor that most clearly defines g_f, CFR, has an average loading of .57 on g_f. This implies that approximately 32% of the variance on the ability represented by CFR is determined by g_f. A similar analysis of *V* indicates that 46% of the variance of the ability represented by that factor is related to g_c. These relatively low loadings of the primaries with the highest average loadings on the g_f and g_c factors indicate that neither g_f nor g_c can be clearly or completely identified with the primaries that are the clearest measures of the factor. Put another way, these loadings are illustrative of the tension between a relatively noninferential and a more inferential interpretation of the factors. In a noninferential sense one may define a factor in terms of the tests or measures that define it and the loadings of the various measures. Thus a factor may be given a neutral name—e.g., a numerical index, and may be defined as that factor on which the following measures have the following loadings. Despite occasional flirtations with such theoretically neutral designations as indicated by Cattell's use of a universal numerical index for factors, most factor analysts have not been content to abjure from more inferential designations of their factors.

Certainly g_f and g_c are given meanings and interpretations that transcend the measures on which they are based. Furthermore, since they are not uniquely defined by existing measures or factors, g_f and g_c are themselves hypothetical constructs with extensive surplus meaning—neither equivalent nor reducible to existing measures nor even well defined by existing measures. In addition to their slightly amorphous empirical anchoring, the interpretation of g_f and g_c as roughly reflecting biological capacity and acculturation is itself highly inferential.

Since the g_f and g_c factors are themselves related, and since the second-order factor solutions used by Cattell are oblique, it is possiable to perform a third-order factor analysis of the second-order factor analysis. In some cases it is possible to perform a fourth-order factor analysis of factors derived from a third-order factor analysis. Since the number of factors derived from a factor analysis is, of necessity, less than the number of variables whose correlations provide the foundation for the factor analysis, it follows that each successive factor analysis will produce a smaller number of more general factors. Thus successive factor analyses will generate a hierarchy as represented in Figure 2.1. Cattell notes that such a hierarchical structure is not to be accepted as an accurate description of the structure of intellect because its presence is mandated by the methodological requirements of successive factor analyses of obliquely rotated factor structures.

Table 2.4 presents a summary of several higher-order (i.e., beyond-second-order factor analyses) factor analyses of ability factors conducted by Cattell and his associates. Since the tests that enter into the factor analysis at the first level are not comparable, the higher-order factors that are derived cannot readily be compared nor can the separate factor analyses be considered replications of each other. Only one factor analysis summarized in Table 2.4, the Horn and Cattell study, includes all of the second-order ability factors that Cattell assumes are present at the second order. And, the results of the higher-order factor analysis in that study are, in some respects, anomalous. What Cattell wishes to emphasize as the most important result of the higher-order factor analyses summarized in Table 2.4 is that, at the third or fourth order, a factor emerges on which g_f loads more highly than g_c, which Cattell designates as $g_{f(h)}$. A second less well-defined factor that emerges as a higher-order factor is called the

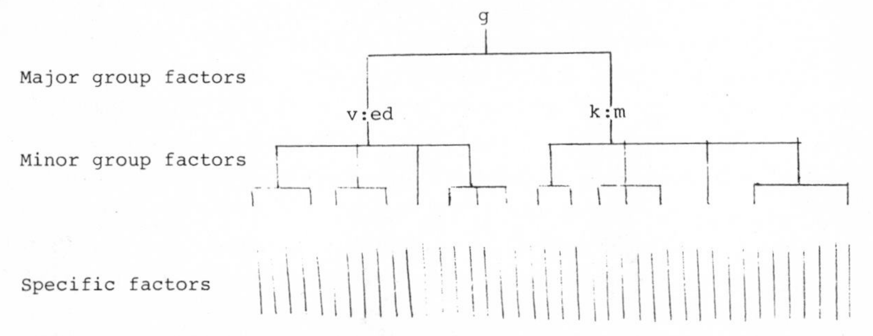

Figure 2.1. *Heirarchical structure of intelligence abilities as generated by successive factor analyses. [Based on Vernon (1961).]*

TABLE 2.4

Cattell's Summary of His Higher-Order Ability Analyses[a]

(A) 5- to 6-year-olds (114)

	$g_f(h)$	Educational effectiveness factor	Possible maturity factor
g_f	.94	−.06	.13
g_c	.41	.38	−.12
Personality factor *X*	.10	.89	.03
Personality factor *Y*	.01	−.01	.93

(B) 9- to 12-year-olds (306)

	$g_f(h)$	Educational effectiveness factor	Possible maturity factor
g_f	.70	−.02	.25
g_c	.59	.48	−.06
Anxiety	−0.9	.07	−.32
Personality factor 1	−.41	.00	.05
Personality factor 2	−.02	.62	.04

(C) 13- to 14-year-olds (277)

	$g_{f(h)}$	Educational effectiveness factor	General personality factor	
			Alpha	Beta
g_f	.69	.00	.02	−.07
g_c	.63	.32	−.04	.07
Anxiety U.I. 24	−.01	.79	−.51	−.07
Exvia U.I. 32	.18	.23	.01	.00
Corteria U.I. 22	.09	.32	−.51	−.07
Personality factor *A*	.01	−.05	.99	−.03
Personality factor *B*	.00	.04	.03	−.74
Personality factor *C*	.02	.06	−.69	.08

(D) Adult Criminals (477)

	$g_{f(h)}$	Educational effectiveness factor	General personality factor	
			Alpha	Gamma
g_f	.53	.02	−.08	−.10
g_c	−.04	.73	.20	−.08
g_r	.42	−.21	−.08	−.40
g_s	.60	−.01	.33	.10
p_v	.57	.38	−.13	.18
Person. U.I.	.00	.11	−.32	−.66
Anxiety U.I. 24	−.03	−.41	−.31	.02
Personality factor *A*	.00	−.01	.45	−.00
Personality factor *D*	.34	.00	.11	.21

[a] Adapted from *Abilities: Their Structure, Growth, and Action*, by Raymond B. Cattell.

educational effectiveness factor. Note that in all four cases reported in Table 2.4 that g_f does not load on that factor. However, in three of the four cases reported, the g_c factor loads on the $g_{f(h)}$ factor. Cattell's interpretation of these findings is in terms of a causal model in which $g_{f(h)}$ is assumed to represent historical fluid ability—that is, fluid ability that was present at an earlier age. Previous fluid ability is assumed to be a determinant of both present fluid ability, g_f, and present g_c. In the three studies of school-age populations, g_c loads more highly on $g_{f(h)}$ than on the educational effectiveness factor. Thus g_c no longer defines a separate mental ability but tends to be subsumed under a factor more clearly related to g_f. The results of the factor analysis of the sample of adult criminals are anomalous in that they indicate that the factor defined as g_c does not load on $g_{f(h)}$. It is not clear whether this is due to the characteristics of the sample of subjects used in the study as Cattell believes (see the explanatory text in Table 2.4) or perhaps to the fact that the higher-order factor analysis was based on a second-order factor analysis that included several ability factors rather than just the g_f and g_c factor. An alternative interpretation of the results reported in Table 2.4 with somewhat different emphases would be to suggest that at the higher order one is able to define clearly only a single ability factor, which is suspiciously reminiscent of Spearman's *g*. The three analyses of the school-age populations clearly support this interpretation. Note that the loadings of the second-order ability factors g_f and g_c are, in each case, substantially higher for the $g_{f(h)}$ factor than for the educational effectiveness factor. As for the study of adult criminals the g_c factor does load separately on the educational effectiveness factor, but with the exception of p_V which has moderate loadings, all of the other ability factors (g_r, g_s, and p_V) load on $g_{f(h)}$. Therefore even for this group, $g_{f(h)}$ seems to be more related to general cognitive ability than does the educational effectiveness factor. Thus, although the results of higher-order factor analysis are tentative, they do point to the existence of a general ability factor which Cattell assumes is the historical precursor of present fluid ability and g_c and which may alternatively be identified with Spearman's *g*.

The causal relationships that exist among the different ability factors have given rise to what Cattell refers to as the investment theory and the triadic theory. The triadic theory involves a distinction between three different types of influences on cognitive performance—*capacities, provincials,* and *agencies.* The theoretical variables that are assumed to represent the most general influences are the capacities. The general capacities are assumed to be related to structural and functional properties of the brain and to influence jointly virtually all cognitive performances. Cattell speculatively suggests that the most general of the capacity factors, g_f, may represent the size of the critical neural substrate for learning. Therefore g_f is given a structural referent. Other capacities, such as g_s and g_r (speed and retrieval capacity, respectively), are speculatively identified with functional characteristics of the brain—e.g., its chemical functioning. Provincials refer to a second class of abilities which reflect powers or abilities of an individual that are not identified with general (either structural or functional) properties of the brain, but rather with powers or abilities that refer to localized brain areas. These provincials have a less restricted influence in test performances. That is, each of the provincials is likely to influence performance in a smaller variety of cognitive tasks than the capacities. The provincials refer to sensory and motor skills that influence performance. The clearest example of such a provincial would be p_V—or visualization ability. Cattell suggests that although species differences in provincials may be marked, the differences within the human species are of somewhat lesser significance, at least within relatively homogeneous cultural groups. Therefore provincial abilities may fill a somewhat lesser role in test performance in the kinds of measures of intelligence used in our culture. The capacities and provincials are determined by genetically influenced characteristics of the nervous system. However, they are both undoubtedly influenced by environmental factors. General capacities might, for example, be influenced by the characteristics of the biological environment and the effects of diet. A provincial such as p_V might be dependent on the kind of visual stimulation experienced by the individual. Also, various insults such as brain damage and strokes ought to affect capacities. An agency, unlike a capacity, is more crucially involved with the cultural experiences of an individual. It comes about through the "investment" of fluid ability and other capacities into a particular intellectual skill which is socially rewarded. The most general agency is g_c which, from the perspective of triadic theory, is given the designation of g to indicate that it represents an agency but it is the only general agency. Other agencies are reflected in somewhat less general intellectual skills but are germane to a narrower class of performances. An example of agencies would be a_V or a_s—symbols of the verbal and spatial agency, respectively. These agencies are identified with the primary abilities delineated in Thurstone's factor analysis. Their designation as agencies reflects their status as theoretical variables within the triadic theory. The agencies presumably are created by the interaction of both general and provincial capacities with the cultural learning experiences of an individual. Thus a_V requires auditory and visual perception of words and may be influenced as well by the more general cognitive capacities. The agencies recognized by Cattell are themselves split into two classes called *aids* or *acquired cognitive skills* which may be thought of as intellectual algorithms that are culturally taught and generally useful, and *proficiencies*, which refer to specialized intellectual skills an individual acquires in order to satisfy some fundamental goal. An example of the latter might be learning a foreign language in order to pursue a scientific career. Proficiencies may occur as idiosyncratic combinations of skills that may be present in relatively small subsets of the population, e.g., the proficiencies involved in being a surgeon.

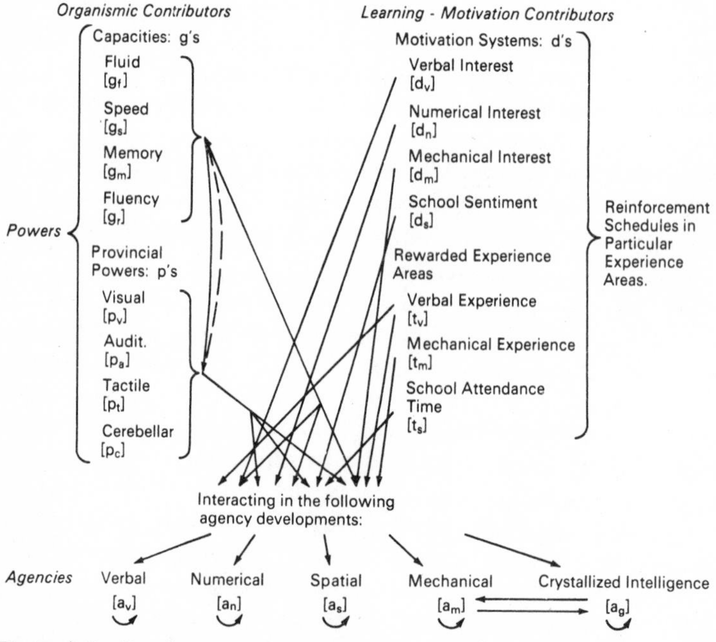

Figure 2.2. *Developmental implications of the triadic theory, worked out consistently with the reticular factor model. The arrows indicate direction of influence and contribution to growth. Thus, verbal ability,* a_V, *receives contributions from the capacities,* g's, *the powers,* p's, *a motivational factor,* d_V, *and a reinforcement in an experience area,* t_V. *To avoid complication of the diagram not all individual but only class connections are made. The semicircular arrows below the agencies indicate their self-development capacities as "aids." [From* Abilities: Their Structure, Growth, and Action *by Raymond B. Cattell. Copyright © 1971 by Houghton Mifflin Company. Used by permission of the publisher.]*

An outline of the triadic theory is presented in Figure 2.2. Note that the figure postulates a number of interacting and causal relations among the various theoretical variables in the theory. This type of structural relationship is called a *reticular factor model* and is distinguished from the hierarchical or "stratum" model. Note in Figure 2.2 that the various agencies are influenced by both the general and provincial powers of an individual. Note further that a_g as a general agency is assumed to be the resultant of specific agencies and to itself act as an influence on other agencies. The general capacities are assumed to influence the provincial capacities and, perhaps more speculatively, the provincial powers may influence the general capacities. Also, the development of agencies may in turn influence general capacities. Finally, a variety of interests and social rewards are assumed to influence the development of agencies.

It is apparent that Cattell's triadic theory is not closely tied to the results of specific factor-analytic investigations. It may be thought of as a speculative attempt to guide research.

In addition to its grounding in the triadic theory, there are a number of other theoretical implications of the distinction between g_f and g_c. Most of these implications arsie from that notion that g_f is more subject to biological influences that affect capacity and g_c (a_g) is more subject to cultural influences. A list of implications follows:

1. Heredity influences g_f more than g_c; but g_f is not equivalent to genetic intellectual capacity. It represents biological capacity, which is dependent upon the influence of the biological environment with respect to such variables as nutrition and prenatal influences as well as genetic endowment. Nevertheless genetic variables affect g_f directly and g_c only in a secondary way through the influence of g_f on g_c. Therefore, we would expect that g_f measures are more influenced by genetic endowment than g_c.

2. Environmental changes that are presumed to affect biological development would have greater influence on g_f than on g_c. Thus we would expect that improvements in nutrition or improvements in obstetrical care would influence g_f directly and g_c indirectly.

3. Environmental changes that affect educational and cultural opportunities would influence g_c but have no influence on g_f at all. Thus, improvements in education should act to change g_c. Similarly, recent attempts to change IQ scores among the disadvantaged by providing special educational opportunities would be expected to show larger influences on g_c than on g_f.

4. Equivalent changes are not shown by g_f and g_c as a function

of age. In particular, measures of g_f are said to show decline at an earlier age than measures of g_c, and g_c measures may show little or no decline at all well into old age. On the other hand, g_f is assumed to start its decline in the third decade of life as the individual's biological efficiency begins its deterioration.

5. Brain damage will have different effects on g_c and g_f. Early in life, brain damage might influence g_f and then influence g_c derivatively. Later in life, brain damage may have relatively little influence on g_c because g_c represents the results of past learning. However, brain damage late in life should have some effect on g_f.

Studies of the effects of brain damage on intellectual ability were the foundation of a biologically based theory of intelligence put forward by Hebb (1942) which is analogous to the g_f–g_c theory. Hebb distinguished between *intelligence A* (potential) and *Intelligence B* (realized intelligence) and tried to show that the physiological evidence, particularly that evidence related to the consequences of brain damage, required such a distinction (Hebb, 1939).

A Preliminary Evaluation of Cattell's Theory

The overall evaluation of Cattell's theory will take place implicitly and explicitly at several points in the book. We shall come to some general statements about factor-analytic models of intelligence at the end of this chapter. Data relevant to the g_f–g_c distinction will be discussed at several other points in the book. For example, we shall discuss data on age changes in intelligence and data on heredity and intelligence in subsequent chapters and we shall at those points deal with their relevance to the g_f–g_c theory. And, we shall return to the theory in our concluding chapter. At this point, in a preliminary evaluation, we wish to deal principally with the data supporting the theory we have presented. In particular we shall deal with three issues and touch briefly on a fourth. These are:

1. The extent to which the factor-analytic studies on which the g_f–g_c theory rests may be considered as replications of each other.
2. Whether the factor structure that supports the distinction is the best or most compelling resolution of the data.
3. The extent to which the causal attributions given in the theory are supported.
4. The extent to which the various implications of the theory are compatible with empirical results.

TABLE 2.5

A More Extensive Research View of Loading Patterns of Fluid and Crystallized Intelligence[a]

(A) 5- to 6-year-olds		
	g_f	g_c
Culture-fair (fluidity markers)	58	−11
Reasoning	10	72
Verbal	−17	74
Numerical	43	49
Personality 2	04	−05
Personality 3	07	−08
Personality C	−07	−09
Personality H	15	17
Personality Q_2	01	02

(B) 9- to 12-year-olds		
	g_f	g_c
Culture-fair (all)	78	09
Reasoning	30	40
Verbal	22	63
Numerical	47	35
Spatial	73	03
Exvia	01	29
Anxiety	05	00
Pathemia	04	04
Neuroticism	−09	06

(C) 13- to 14-year-olds		
	g_f	g_c
Culture-fair (classification)	63	−02
Reasoning	08	50
Verbal	15	46
Numerical	05	59
Spatial	32	14
Personality F	−05	09
Personality C	21	−07
Personality H	21	−04
Personality Q_2	−06	05
Personality Q_3	05	−02

(D) Adults		
	g_f	g_c
Culture-fair (all)	48	−08
Reasoning	26	30
Verbal	08	69
Numerical	20	29
Spatial	04	−04
Mechanical knowledge	−15	48
Speed of perceptual closure	18	−05
Ideational fluency	−03	25
Inductive reasoning	55	12
Personality, U.I. 16	−04	18
Personality, U.I. 19	05	07
Personality, U.I. 21	−03	−08
Personality, U.I. 36	01	43
Personality, anxiety, U.I. 24	−05	−26

[a]Adapted from *Abilities: Their Structure, Growth and Action* by Raymond B. Cattell. Copyright © 1971 by Houghton Mifflin Company. Used by permission of the publisher.

1. THE REPLICABILITY OF THE FACTOR STRUCTURE

We have already noted that there are some differences in the third- and fourth-order factor loadings leading to the derivation of $g_{f(h)}$ in Cattell's studies. Table 2.5 presents a summary of the several second-order analyses published by Cattell in support of his theory. An examination of Table 2.5 indicates that there are both consistencies and inconsistencies among the several studies. First, the consistencies. The culture-fair tests load positively on g_f and negligibly on g_c. Verbal tests that serve as markers for the verbal ability factor always load positively on g_c and neglibibly on g_f. Inconsistencies are present in the loadings of the spatial, reasoning, and numerical tests. Spatial tests substantially load g_f (they load g_f about as highly as the culture-fair tests) in Study B with 9- to 12-year-olds and have no loading on g_c in this study. In Study C, spatial tests have a somewhat lower relationship to g_f and a marginally higher relationship to g_c. In Study D, with adults, spatial tests have essentially zero loadings on both g_f and g_c. Reasoning tests load g_c but not g_f in Studies A and C; but in Studies B and D, reasoning tests have lower loadings on g_c which are almost equal to their loadings on g_f. Numerical tests load both g_f and g_c moderately high in Studies A and B. They load g_c high and g_f low in Study C and they load both g_f and g_c with low values in Study D. We are, of course, dealing with different age groups and different tests to index the various abilities. Nevertheless, the assertion that the distinction between g_f and g_c is well established and replicable demands that in the several studies supporting the distinction the same results have been established. Given the inconsistencies in factor loadings we cannot assert that the separate g_f and g_c factors that emerged are identical. Of course some of the variations in g_f and g_c at different age levels might be related to the theoretical notion that g_f influences future g_c. In this connection Cattell suggests that numerical ability may reflect fluid ability in 5- and 6-year-olds but at a later age it may be more related to computational skills taught in the school. However, this sort of reasonable, although probably ad hoc, explanation will not do for all of the inconsistencies we have noted in Table 2.4. For example, it is not at all clear why spatial ability should show such large changes in loadings. We may conclude therefore that the distinction between g_f and g_c requires further empirical elaboration and replication in a variety of subject populations and using a variety of tests. What does appear as well established in the four studies summarized in Table 2.5 is that tests of verbal ability emphasizing vocabulary apparently load a different factor than the Cattell culture-fair tests of ability in second-order factor analyses.

2. IS THE FACTOR SOLUTION FAVORED BY CATTELL OPTIONAL OR COMPELLING?

Humphreys (1967) has published a critique of one of the Cattell studies on which the g_f–g_c distinction rests (Cattell, 1963). Humphreys criticizes the factor-analytic procedures used in the study. Some of the technical aspects of the criticism are beyond the scope of our discussion. One point is fairly clear. Cattell included a number of personality measures in the factor analyses in order to provide "hyperplane stuff" or variables that contrast with the ability measures. Humphreys argues that these variables add noise to the analysis and obscure the results. When these are omitted, nine ability measures remain. A factor analysis of these measures yields the factor solutions presented in Table 2.6.

Humphreys favors the factor solution based on two factors since these seem to account for most of the variance in the matrix. In this connection note that h^2, which represents the communalities or the amount of variance accounted for in the various tests by the respective factors, does not substantially increase in most cases if one adds Factors III and IV. With the two-factor solutions favored by Humphrey, the factors are correlated .57 (as opposed to .44 in Cattell's original analysis for the g_f and g_c factors). In Humphreys's analysis the first factor includes both the culture-fair tests (variables 6 to 9) and the primary mental abilities (PMA) defined by Thurstone. Thus the factor is more pervasive than g_f and is interpreted by Humphreys as an intellectual-educational factor. Thus the first and most substantial factor appears to cut across the g_f and g_c factors. The second factor is perhaps closer to Cattell's g_f in that the culture-fair tests and spatial ability load positively on it whereas the other primary mental abilities are negatively related to it. Again note that the substantial correlation between the first and second factors derived by Humphreys ($r = .57$) justifies a higher-order analysis in which one general factor emerges. The analysis performed by Humphreys while partially supporting Cattell in indicating that a factor can be derived for which alleged measures of fluid ability are not related to other primary abilities, also supports the type of analysis favored by British factor analysts such as Vernon (1961). British factor analysts perform their factor analyses by first extracting a g factor, which, in samples with a representative range of talent, i.e., in samples that have not been selected for intellectual ability, might account for as much as 50% of the variance in the matrix. After accounting for g, the British factor analysts typically extract two additional factors, $v{:}ed$ which represents verbal-numerical and educational abilities and $k{:}m$ which represents practical-mechanical and spatial abilities. A typical British factor solution based on the performance of 1000 army recruits is presented in Table 2.7.

TABLE 2.6

Humphreys' Reanalysis of a Study by Cattell: Two-, Three-, and Four-Factor Solutions for the Ability Variables[a]

	Two			Three				Four				
Variable	I	II	h^2	I	II	III	h^2	I	II	III	IV	h^2
1	.64	−.20	.45	.65	−.22	−.14	.49	.64	−.20	−.13	.10	.48
2	.48	.15	.25	.50	.17	−.32	.38	.49	.17	−.29	.14	.37
3	.63	−.19	.43	.63	−.19	.02	.43	.64	−.19	−.08	−.24	.51
4	.54	−.34	.40	.54	−.34	.05	.41	.54	−.33	0	−.15	.42
5	.57	−.33	.43	.57	−.34	.06	.44	.59	−.38	.15	.21	.55
6	.56	.20	.35	.55	.19	.06	.35	.56	.21	.15	.16	.41
7	.51	.35	.38	.50	.33	.01	.36	.50	.34	−.01	−.06	.37
8	.65	.24	.48	.67	.27	.27	.60	.66	.27	.24	−.11	.58
9	.44	.24	.25	.44	.23	−.11	.26	.44	.24	−.12	−.01	.26

[a] Based on Humphreys (1967).

TABLE 2.7

Simple Summation and Group Factor Analyses of Tests Given Given to 1000 Army Recruits[a]

Tests	Group factors					h^2
	g	*k:m*	*ed.*	*v*	*n*	
0 Progressive matrices	.79	.17				.65
Dominoes (nonverbal)	.87					.75
Group test 70, Pt. I	.78	.13				.62
4 Squares	.59	.44				.54
3 Assembly	.24	.89				.85
2 Bennett mechanical	.66	.31				.54
25 Verbal dictation	.79		.29	.45		.90
14 A.T.S. spelling	.62		.54	.48		.90
21 Instructions	.68		.41	.43		.82
	.87		.23	.09		.82
3A Arithmetic, pt. I	.72		.49		.39	.91
Arithmetic, pt. II	.80		.38		.16	.82
23 A.T.S. arithmetic	.77		.36		.32	.82
Variance (%)	52.5		8.7	8.4	6.9	76.2

[a] Based on Vernon, P.E. *The Structure of Human Abilities*, New York, Wiley, 1961.

Note in Table 2.7 that *g* accounts for 52.5% of the variance and that all of the tests show some loading on *g*. The best definition of *k:m* is by tests of manual dexterity and mechanical ability, and *v:ed* is defined by a variety of tests reflecting educational, verbal, and numerical skills. There is some support for Cattell's position in these data. Note that the Progressive Matrices test, a nonverbal measure of intelligence that involves spatial analogies and is related to the Cattell culture-fair items, has no loading on the *v:ed* factor which is analogous to g_c.

Humphreys (1967) notes that the Horn and Cattell study is in disagreement with Vernon's analysis. Even though, Humphreys argues that an optimal factor-analytic solution of the study by Horn and Cattell (1967) using adult criminals would produce three ability factors—a g_f factor which combines g_v or visualization ability, g_c, and a factor related to speed and fluency. However, the test representing mechanical ability is somewhat more closely related to verbal ability than to spatial ability in the Horn and Cattell study.

If we consider jointly the results of Cattell's factor analyses, Humphreys' reanalysis and critique of these results, and the factor solution favored by the British school, we find some broad areas of agreement and some disagreements. Perhaps most fundamental of all is that there is a pervasive influence of a single general ability, *g*, in tests of intellectual ability. Whether *g* is present in all or only a substantial subset of such tests is perhaps an indeterminate question. That *g* accounts for a substantial portion of the variance is clear.

These analyses also are in agreement that *g* is not the only variable that must be postulated to account for all of the variance. Thus, the tetrad difference law implied by Spearman's original theory is not correct. There is somewhat less agreement with respect to the additional narrower factors that must be postulated. To a rough extent there is agreement that certain skills closely related to academic subject matter are separable from the kinds of things measured by culture-fair intelligence tests. However, the precise location of some of the primary abilities in this scheme is not always clear. Thus the relationship between spatial and mechanical abilities is not invariant over all of the analyses. Some of the disagreements are due to difference in the methods of analysis used, differences in the tests used to define the factors, and differences in the samples of subjects used in the various studies.

3. THE CAUSAL ASPECTS OF CATTELL'S THEORY

Cattell's theory assumes a rather complex set of causal relationships among the various ability factors. It is apparent that many of the causal interpretations made by Cattell are, at best, only weakly required by the factor-analytic results themselves. One can however treat such causal attributions as hypotheses that can be tested. Such hypotheses will probably require longitudinal designs for critical evaluation. Consider, for example, the hypothesis that g_f at a particular time is an antecedent for subsequent g_f and g_c, whereas g_c at a particular time does not influence subsequent g_f. Such hypotheses can be tested by means of cross-lag panel correlations. Let us illustrate this. Suppose we have for the same group of individuals measures of g_f and g_c at Time 1 (t_1) and measures of g_f and g_c at a later time (t_2). Assume further for simplicity that the correlation between g_f and g_c at t_1 is equal to their correlation at t_2. Assume further that the correlation between g_f at t_1 and g_f at t_2 is equal to the correlation between g_c at t_1 and g_c at t_2. Then, if g_f influences subsequent g_c but g_c does not influence g_f, we would expect that the correlation between g_f at t_1 and g_c at t_2 is greater than the correlation between g_c at t_1 and g_f at t_2. Cattell has not performed such studies and therefore the critical data required for an evaluation of the causal relationships between g_f and g_c have not been obtained. Thus, the causal relationships in the theory have the same status as a number of the other implications that are suggested by the g_f–g_c distinction—they remain to be critically tested in subsequent research.

J. P. Guilford

Guilford's theory of intelligence represents a radical departure from the Spearman-Thurstone tradition. (For general presentation of his views, see Guilford, 1967, and Guilford and Hoepfner, 1971.) Perhaps his most significant point of contact with that tradition is in the use of factor analysis as a means of discovering the structure of intellect. However, Guilford tends to use factor analysis as a means of testing a hypothetico-deductive model. Thus factor analysis is used as a means of confirming a structural model rather than as a means of discovering a structural model. Also, Guilford carries Thurstone's criticism of *g* much further. He assumes, as we shall see, that there are 120 separate types of intellectual abilities. And, he does not accept the notion that *g* can be derived as a higher-order factor.

Guilford's model is based on the notion that there are dimensions whose combinations determine different types of intellectual abilities. One dimension of an ability is the kind of mental operation involved in the ability. Guilford distinguishes five types of mental operations. These are cognition (knowing), memory, divergent production (generation of logical alternatives), convergent production (generation of logic-tight conclusions), and evaluation. The second dimension of classification is in terms of content or areas of information in which the operations are performed. He distinguishes four types of content—figural, symbolic, semantic, and behavioral. The third dimension of abilities is the product that results from a particular kind of mental operation applied to a particular type of content. Guilford distinguishes six types of products. There are units, classes, relations, sytems, transformations, and implications. If we consider all possible combinations of operations, contents, and products we find that there are 120 different abilities that may be defined by this structure of intellect model. Figure 2.3 is a graphic representation of the model and Table 2.8 presents the code used by Guilford to describe the various abilities that are assumed to exist according to the model.

Note that according to the model, each ability is defined by its unique position on each of three dimensions. It is not assumed that abilities that share position with respect to two dimensions but differ in a third are necessarily more related than abilities that share only a single dimension. Put another way, Guilford does not assume that the dimensions of the model are higher-order factors. If Guilford's

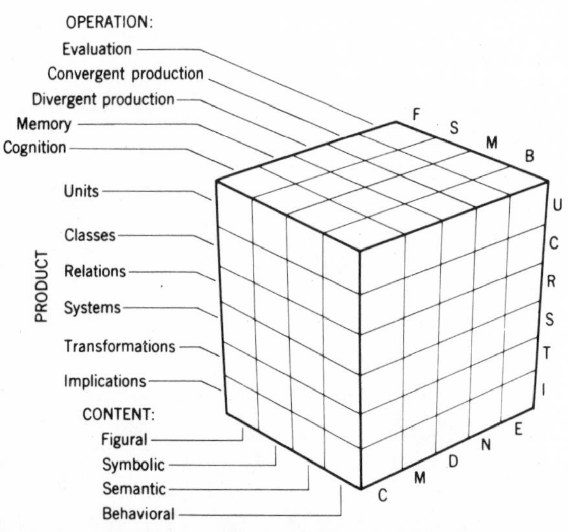

Figure 2.3. *Guilford's model of intelligence. [From Guilford, J.P.* **The Nature of Human Intelligence.** *New York: McGraw-Hill, 1967.]*

TABLE 2.8
Symbols Used in Guilford's Model

Operation	Content	Product
C—cognition	F—figural	U—unit
M—memory	S—symbolic	C—class
D—divergent production	M—semantic	R—relation
N—convergent production	B—behavioral	S—system
E—evaluation		T—transformation
		I—implication

analysis is correct it should be theoretically possible to construct tests that are defined by a single ability. In terms of factor-analytic methodology it should be possible to construct tests that load substantially on only one of the many different factors that can be extracted in a single study.

In order to give the reader a clearer idea of the nature of the abilities postulated in Guilford's structure of intellect model, we will provide examples of several abilities and the measures that define them.

CFV—Cognition of Figural Units

An example of the type of test that would measure this ability is "Hidden Print." The subject is presented pictures of digits and letters formed by patterns of dots. Ther are in addition a variety of dots scattered at random about the dots forming the digits and letters which serve to obscure the stimuli and make correct recognition difficult. The subject is required to recognize the stimuli presented to him. The task involves cognition because the subject is required to become aware or discover something. The stimuli or content that requires this operation is presented in the visual mode and is therefore figural. (The term *figural* also includes content presented in other modalities.) Finally the product of the operation is a unit, in this case a particular digit or letter.

MSC—Memory for Symbolic Classes

An example of a test assumed to measure this ability is the memory for name and word classes. The subject is presented a set of names such as IRIS, IRENE, IRVING. The subject is then required to determine in a recognition test whether certain words do or do not belong to the class presented to him. The subject might be presented the word IRA and IDA and would be required to recognize that IRA belongs to the class but IDA does not. The test clearly involves the operation of memory. The stimulus the subject deals with is symbolic in form and the product clearly is a class concept.

DMR—Divergent Production of Semantic Relations

This ability is defined by tests of associational fluency. An example would be the presentation of the pair of words FATHER–DAUGHTER and the subject would be required to list all of the ways in which the pair are related, e.g., parent–child, old–young, and male–female. Divergent production tests are scored in terms of the number of acceptable answers. Such a test is clearly divergent (there is no single correct answer). The content of the item is clearly semantic, and the product deals with a relation between the contents.

CBI—Cognition of Behavioral Implications

A test called "Reflections" is assumed to measure this ability. In this test, a subject is given a statement of the kind that a patient might make during psychotherapy. The subject is required to pick the best psychological implication of the statement. For example, the statement might be, *I'm just wondering how I'll act; I mean how things will turn out.* The three alternative implications might be:

1. *She's looking forward to it.*
2. *She's worried about it.*
3. *She's interested in how things will turn out.*

The second answer is supposed to be correct. The test deals with behavioral (psychological) content and the product is clearly an implication.

EMT—Evaluation of Semantic Transformations

A test that is said to involve this ability is "Story Titles." In this test the subject is presented a brief story and several possible titles. The subject must suggest the title that is best in terms of the story. The test involves an evaluative judgment with respect to semantic content. And the product of the evaluation results in a transformation.

There are in all 98 different factors (and tests defining them) out os a possible 120 that have been identified by Guilford in the various studies he has undertaken. (Descriptions of tests measuring the various factors that have been defined can be found in Guilford, 1967, and Guilford and Hoepfner, 1971.)

An Evaluation of Guilford's Model

Unlike Spearman, Thurstone, Vernon, Burt, and Cattell, Guilford does not deal with higher-order or more general factors. He insists that the factors discovered in the structure-of-intellect model are found with orthogonal rotations and hence are independent of one another. Therefore, in evaluating Guilford's model one must come to grips with its radical rejection of theories that emphasize general ability factors. One also must reconcile Guilford's data with the data we have reviewed in our evaluation of Cattell's theory which led to the conclusion that there was a general intellectual factor. Guilford (1964) has pointed out that 17% of all the correlations among tests of intellectual abilities in his earlier studies fell within the interval from −.10 to +.10. In his 1971 summary of the status of his theory, he indicated that the percentage of correlations in the interval between −.10 and +.10 was 18 (8677 of 48,140 coefficients). And, for 24% of the correlations found in his numerous studies one could not reject the null hypothesis that $r = 0$. These data, Guilford argues, simply do not support the view that there exists a single pervasive *g* factor in intellectual ability.

Even if we expect Guilford's findings at face value, we find that the most comprehensive research on the most diverse intellectual abilities and measures specially constructed and selected to be independent of each other still leaves us with the finding that 76% of the time we can reject the null hypothesis that $r_{ab} = 0$ (where a and b represent any pair of ability measures) and infer that $r_{ab} > 0$. Strictly speaking, this finding is not compatible with Guilford's theory. If measures of intellectual ability are unrelated except if they are measures of the same ability we would expect measures selected at random from Guilford's investigations to correlate positively with relative infrequence.

In a typical investigation Guilford is able to derive approximately 20 factors using approximately 50 variables. Assume that it is the case that the typical variable in his investigation is not a pure measure of ability but does in fact load on two of the 20 ability factors present. Then the probability that two such variables will load on at least one common value is considerably less than .76. Therefore, it is difficult to see how on Guilford's model one could account for the high number of correlations that are significantly greater than zero. In order to explain that finding, one would have to assume that factors more general than those extracted by Guilford exist in the matrices of correlations he deals with.

In addition, for a number of reasons, the figure of 24% of correlations not significantly different from zero needs to be accepted with caution. First, many of the variables dealt with in Guilford's studies have relatively low reliabilities—sometimes reliabilities below .50. Such variables would not be expected to correlate highly with other variables because substantial portions of their variance are error variance. Recall that Spearman pointed out in 1904 that the influence of *g* could be obscured where there was a failure to correct for attenuation. Second, in a number of early studies, Guilford used Air Force cadets in an officers' training program. His subjects were selected both for intellectual ability and for special aptitudes that were germane to flying ability. Therefore, the sample used was biased. Any restriction in the range of talent will decrease the value of correlations between variables. In effect, Guilford often dealt with samples where some of the influence of *g* had been eliminated. In some of his later studies the samples used were high school and junior high school students where ability selection had not occurred. And, for evaluating the pervasiveness of *g* such sample are preferable. Third, Guilford's measures include areas of ability that stretch the meaning of the term *intellectual ability*. Two particular types of abilities come to mind—those involving behavioral content and those involving the operation of divergent thought.

The behavioral content area deals with sensitivity to psychological states and feelings. It may be the case that this type of ability is unrelated to more general intellectual abilities. Also, there is little evidence that these measures have predictive validity. Hence, we are forced to accept on the basis of face validity (that is, on what appears obvious on inspection) that the measures are indeed measures of abilities germane to the prediction of psychological characteristics. A similar question can be raised with the divergent thinking category. Traditionally, measures of intellectual ability have had a single correct answer. Divergent thinking measures are scored in terms of the number of different acceptable answers suggested. Guilford believes that a number of these measures are related to creativity. This is an issue we shall discuss in a subsequent chapter. It should be noted that there is some question whether divergent thinking measures are measures of intellectual ability. Parenthetically, it should be noted that there is some question whether they are measures of creativity. Indeed, there is some question whether they are measures of anything other than divergent thinking ability at all. To raise this question is, by implication, to point to an ambiguity in the notion of *g*—what is a measure of intellectual ability? To say that all measures of intellectual ability measure are, to some extent, the same thing is meaningless unless one knows what constitutes an intellectual ability. Thus, to say that Guilford may include variables that are not intellectual ability measures may only be a way of saying that we have conceived the notion of an intellectual ability in an excessively narrow and traditional manner.

In any case the designation of new tests as measures of abilities must be accepted cautiously until the validity of the tests is established. We know that more traditional measures of ability have predictive validity in that they may be used to predict, inter alia, academic success. We have very little evidence that divergent thinking measures or measures dealing with behavioral content are valid measures in the sense that they may be used to predict performance in real-life situations.

We have argued that Guilford's figure of 24% is unrealistically high. In order to make this point more concrete we can consider two later studies from Guilford's laboratory that deal with traditional intellectual abilities and use samples of subjects that are reasonably representative of general ability (Tenopyr, Guilford, & Hoepfner, 1966; Dunham, Guilford, & Hoepfner, 1966). Tenopyr *et al.* (1966) have reported a factor analysis of symbolic memory abilities that included 50 tests and resulted in the definition of 18 factors define in the structure-of-intellect model. In examining their data we excluded all tests whose reliabilities were below .6. This removed seven variables from consideration. There were 903 correlations that were reported among the remaining 43 variables. For the size sample in their study, any correlation with an algebraic value of .12 or less would not be sufficiently large to reject the null hypothesis that $r = 0$. There were 19 such correlations, or approximately 2% of the correlations reported, that were not significantly different from zero. Note also that by chance one would expect that some correlations would be less than .12 even where their true value in the population was .12 or more. One other finding in this study is worth noting. One of the tests included in the battery was the SCAT, a standardized test of intellectual ability that is probably a good measure of g_c. The SCAT correlates between .16 and .70 with all of the remaining variables in the battery (excluding those with reliabilities below .60). The median correlation was .39. Thus a standardized test used in schools could be used to predict performance on every other test in the battery. Corrected for attenuation, one could estimate that approximately 25% of the variance on all measures in the battery is known given an individual's score on the SCAT—and this occurs in a battery where Guilford's factor solution involves the postulation of 18 orthogonal factors.

These results are not totally anomalous. Dunham, Guilford, and Hoepfner (1966) have reported the results of a factor analysis of abilities pertaining to classes and the learning of concepts. Again, a sample with something approximating a representative range of ability was used. Included were 43 different tests, and 15 different orthogonal factors were derived as the final factor solution. Again, excluding variables with low reliabilities we find that less than 1% of the correlations are not significantly different from zero. One of the measures was a multiple-choice vocabulary test—again a good measure of g_c in Cattell's system. Its correlation with the remaining variables in the matrix, excluding those with low reliabilities (below .5) ranges from .27 to .60 with a median value of .45, uncorrected for attenuation. This again indicates that a substantial portion of the variance in all of the measures used in this investigation is predictable by knowledge of scores on a multiple-choice vocabulary test.

We can conclude that Guilford has not really demonstrated that broad general ability factors do not exist, nor indeed has he proved wrong the notion that there exists one pervasive ability factor that enters into a great variety (if not all) intellectual performance. If this conclusion is correct then Guilford's structure-of-intellect model cannot be a completely accurate or exclusively correct analysis of the structure of intellectual abilities. However, this still leaves open the question of the meaning of the various ability factors that have been defined in Guilford's research. We shall deal with this question in three ways. First, we shall discuss evidence with respect to the replicability of the factors derived by Guilford. Second, we shall consider the extent to which the factor structures postulated by Guilford support the structure-of-intellect model. Third, we shall deal with the predictive validity of scores on the various factors derived in Guilford's research.

REPLICABILITY OF FACTOR STRUCTURES

Replicability refers to the tendency to be able to reproduce a set of empirical results. With respect to factor analysis, there are a number of different kinds of replicability. These include the ability to obtain the same set of factors using the same tests with the same or different samples of subjects. With this type of replicability, numerical indices exist that permit one to measure the extent to which a factor structure has been replicated. In such a case replication refers to the ability to reproduce an identical set of test loadings on the same set of factors. Other types of factor replicability exist. For example, we might use a different set of tests with the same or different samples of subjects where the tests used in the replication study are structurally or logically analogous to those used in the original study. One would then attempt to obtain a similar factor structure in the replication study. Numerical indices of replicability are not really appropriate here because one cannot ask if the same test had the same loadings over the same set of factors. In most of the investigations conducted by Guilford and his associates, new tests are used. Hence, numerical indices of replicability are not applicable. Rather, in most instances replicability consists of an intuitive judgment that a factor derived in a particular investigation is the same or has the same meaning as one derived in a previous investigation. Such a judgment is based typically on the observation that a particular test, thought to be a particularly good measure of one of the abilities and taken as a "marker" variable for the factor, is found to have a high loading on the allegedly invariant factor found in two different investigations. However, since the exact battery of tests used in two such investigations are rarely identical in Guilford's research, one must infer that those tests that define the factor in one investigation, and are not included in a second study in which the factor is allegedly derived, would similarly define that factor. Thus, an element of conjecture is involved in the subjective judgments of factor invariance used in Guilford's research. In the few cases where numerical indices of invariance could be computed in Guilford's research program, "their uses led to unimpressive results [Guilford & Hoepfner, 1971, p. 42]." The fact that numerical objective indices of replicability are not satisfactory leads one to believe that the subjective judgments of factor replicability are suspect.

Guilford's demonstrations of factor replicability are based on "targeted rotations." A targeted rotation is one in which there is an attempt to guide the factor solution by hypothesizing an ideal set of loadings for each of the tests in the sample. Such a factor solution can be contrasted with one that uses an analytic procedure such as varimax in which there is no attempt to guide the results of the solution by a priori notions. Guilford indicates that it is possible to obtain satisfactory factor solutions (e.g., solutions satisfying simple structure) either by the use of targeted solutions or through varimax rotations. Only 32% of the factors obtained by the use of varimax rotations could be satisfactorily identified with factors postulated in the structure-of-intellect model. The corresponding percentage for targeted solutions was 92 (Guilford & Hoepfner, 1971, p. 55). This dramatic difference in percentages has implications for understanding the extent of factor replicability in Guilford's research. The fact that only 32% of the factors derived using varimax rotations are identifiable with theoretically defined factors indicates that such factors are not immediately present or apparent if an unbiased examination of the obtained correlation matrix is undertaken. The fact that solutions can be obtained that are in line with theoretical expectations if targeted solutions are used suggests that Procrustean efforts are required to obtain satisfactory results. Such obtained agreements with theoretical expectations are far less impressive than the demonstration of empirical agreement with theoretical expectations obtained without efforts to manipulate the empirical results in order to present them in the most favorable light.

THE CLARITY OF FACTOR STRUCTURES

The structure-of-intellect model implies an ideal factor structure. In an investigation involving tests representing a variety of abilities postulated by the model with at least two tests chosen to represent each ability, we should find that the tests representing a factor should have substantial loadings on that factor. If tests do not load substantially on a particular factor then the factor becomes somewhat devoid of empirical content since it cannot be identified clearly with the common content of a set of tests. And, if tests load substantially on more than one factor, then the goal of precisely identifying abilities so that pure tests of each ability can be constructed may be said to be unattained.

Guilford and Hoepfner (1971) have reported the results of a reanalysis of all data in Guilford's research program using common analysis procedures and the benefit of hindsight to provide theoretically optimal factor solutions for all of the tests in their study. For each of the abilities discovered at that time, they report the tests that load on the factor with the loadings of each of the tests in every investigation in which it was related to that factor. Table 2.9 presents a summary of all of these analyses. The right-hand column in Table 2.9 presents the lower value of the two tests with the highest factor loadings in any investigation. Note that the median lower bound is in the .40 to .49 range. Very few of the factors are defined by two tests in an investigation with loadings of .60 or more. These figures exaggerate the loadings of the tests on the factor since they do not allow for sampling error. That is, a test may have a spuriously high loading in a particular investigation that is not replicable. A somewhat more conservative summary of the typical loadings is given in the left-hand column of Table 2.9. This column presents the lower value for the two tests with the highest replicated loadings on the factor. In this case a test's lowest replicated loading is defined as the lowest value obtained in all of the investigations in which that test defined the factor. In some cases that value may be the lowest of two obtained loadings. In other cases it might represent the lowest value obtained in five or six investigations. This summary is overly conservative since a test that has repeatedly loaded on a factor in several

TABLE 2.9
Frequency Distribution of Loadings of the Two Tests with the Highest Loading on Each of 99 Factors in the Guilford Model[a,b]

Loadings	Replicated[c] $f(x)$	Nonreplicated $f(x)$
.70–.79	0	7
.60–.69	1	22
.50–.59	6	36
.40–.49	23	31
.30–.39	26	3
Not replicated	43	—
	99	99

[a]The loading for the factor with the second-highest rating is tabulated.

[b]Singlets are not included.

[c]The replicated value refers to the lowest value obtained in all studies in which the test loaded on the factor. The tabled value represents the lowest loading of the test with the second highest loading on the factor.

studies may have a spuriously low loading due to sampling error in a single study. Nevertheless the summary indicates a very substantial lack of replicability for the abilities defined in Guilford's model. For 43 of the abilities there do not exist two studies in which the same two tests defined a factor. And, where replication does exist, the lower value of the two tests that define the factor is almost invariably in the .30 to .49 range (or, in terms of variance accounted for by this factor on the test, in roughly the 9% to 25% range).

Table 2.10 presents a similar analysis for the single test with the highest loading on each of the defined factors. The right-hand column presents a frequency distribution of the loadings of the test with the highest loading in any study for each of the factors. Slightly more than half of these tests have loadings .60 or more. The left-hand column of Table 2.10 presents the highest replicated loading achieved by any test in the studies where the test defines the factor. Note that the loadings of a test on a factor have not been replicated in a substantial percentage of cases. And, where the tests are replicated, the great majority have loadings of .59 or less.

The data summarized in Tables 2.9 and 2.10 indicate that Guilford's factors are not well defined by existing measures. A relatively lenient standard for adequate definition of a factor might be that two tests load .60 or more on the factor in two or more studies. Only one factor defined by Guilford can meet this standard. In fact only

TABLE 2.10
Frequency Distributions of Highest Loading on Each of 99 Factors in the Guilford Model[a]

Loadings	Replicated loading[b] $f(x)$	Not replicated $f(x)$
.80–.89	0	2
.70–.79	0	16
.60–.69	4	35
.50–.59	18	37
.40–.49	26	9
.30–.39	15	0
Not replicated	36	—
	99	99

[a] Singlets are excluded.

[b] The replicated loading refers to the lowest value obtained in separate replications for the test with the highest of these values.

seven of the abilities meet the standard of two or more tests with loadings no lower than .50 in all of the studies in which the tests appear. These data help us to explain why the results of numerical analyses of replicability are not impressive in Guilford's research. Guilford's studies indicate that weakly defined orthogonal factors can be derived to support his theoretical assumptions. However, greater clarity of factor structure is dependent upon the selection of tests that have consistently high loadings on a single factor. Guilford has yet to demonstrate the existence of such tests for the vast majority of abilities defined in the structure-of-intellect model.

Cronbach (1970a,b) has dealt with the clarity of factor structures in Guilford's research (see also Guilford, 1972, for a reply to Cronbach and to other similar criticisms). Cronbach has indicated that the correlations between tests defining the same factor are not substantially different from tests defining different factors. Cronbach has also noted that tests with the same content tended to be more clearly related than tests with different content. However, tests with the same content but different products were difficult to differentiate in the sense that a test with a moderately high loading on a particular factor is likely to have a high loading on a factor with the same content but different product.

This analysis led Cronbach to suggest that there are probably broad general factors present in the matrices analyzed by Guilford that are obscured by his analyses. Guilford has responded by agreeing that it is in fact the case that tests defining different content factors are easier to differentiate than tests defining different product factors. Guilford attributes this to the notion that tests are not univocal in his analyses. That is, tests do not typically load on a single factor. Therefore, two tests that define different factors may correlate with each other because they share common variance on a number of factors other than those they are said to define. Guilford goes on to indicate that this situation can be rectified by refinements in test construction so that tests become univocal. However, this is a promissory note. Until such tests exist, it is possible to assert that the factors that define the structure-of-intellect model have not been clearly differentiated.

PREDICTION AND THE STRUCTURE-OF-INTELLECT MODEL

Even though we have argued that the clarity of the factor structures in Guilford's studies leaves something to be desired, it might still be the case that the various abilities defined in this research are useful for the purpose of predicting performance. What we wish to know is the extent to which knowledge of an individual's score on a variety of abilities enables us to make predictions about that individual that are superior to those that could be made on the basis of some more global index of intellectual ability. If it could be shown that substantial improvement in prediction in a variety of important contexts could be obtained by knowledge of an individual's scores in a variety of abilities in the structure-of-intellect model, then we would have evidence for the utility of the structure postulated, even though it may be imperfectly defined at present. There are several studies available that deal with this issue. Guilford and Hoepfner (1971) present a number of studies relating scores on tests (or factors) representing a variety of abilities to various external criteria. One of the studies reported by them is particularly germane to our purposes. It involves a comparison of the predictive validity of scores on tests representing the various structure-of-intellect abilities with the predictive validity of scores on standardized tests of academic aptitude. The study reports an attempt to predict grades in Grade 9 arithmetic courses. One test was selected to represent each of 25 different abilities that previous research had suggested might be relevant to arithmetic ability. The criteria to be predicted were grades in four different arithmetic courses offered to Grade 9 students. In addition, scores on standardized group tests of academic ability were available for comparison purposes. Table 2.11 presents

TABLE 2.11
Multiple Correlations for Predictions of Mathematical-Achievement Scores from Weighted Combinations of Standard Tests and of Factor Tests[a,b]

Prediction composite	Basic mathematics	Noncollege algebra	Regular algebra	Accelerated algebra
9 Standard tests	.60	.53	.22	.74
2 CTMM scores	.34	.40	.18	.37
3 Iowa tests	.53	.31	.20	.62
4 DAT tests	.57	.53	.24	.70
7 Factor tests	.42	.56	.27	.51
13 Factor scores	.46	.45	.39	.75
20 Factor predictors	.48	.54	.38	.74

[a] Based on Guilford and Hoepfner (1971).
[b] The multiple *r*s are unbiased, i.e., corrected for shrinkage.

the results of a series of multiple correlations in which the prediction of grades involving an optimal combination of scores are compared for standardized tests and for tests relating to factors in Guilford's model. Table 2.11 indicates that for two of the four courses (noncollege algebra, accelerated algebra), standardized tests are about as accurate in prediction as predictions derived from factor scores. For one of the courses, basic mathematics, the standardized tests are superior and for another of the courses, regular algebra, the factor predictors are superior. There is little evidence in these data that the abilities in the structure-of-intellect model are particularly useful for the prediction of grades in mathematics.

Another analysis of the data in the same study is presented in Table 2.12. Table 2.12 indicates whether the addition of factor scores in the prediction equation adds to the ability to predict grades after scores on standardized tests are known and taken into account. That is, the analysis reported in Table 2.12 may be taken to indicate whether or not knowledge of an individual's score on tests representing various abilities yields predictively useful information not contained in scores on tests representing more global ability measures.

TABLE 2.12
Increases in Multiple Correlations (r) from Adding Thirteen Factor Scores to Each of Three Standard Composites from Academic-Aptitude Tests, and F Ratios for Testing Significance of Increases[a,b]

	Basic mathematics		Noncollege algebra		Regular algebra		Accelerated algebra	
	r	*F*	*r*	*F*	*r*	*F*	*r*	*F*
CTMM (2 scores)	.35		.41		.21		.38	
CTMM+13 scores	.59	1.58	.59	1.60	.54	2.25[c]	.80	6.06[d]
Iowa (3 scores)	.55		.34		.24		.63	
Iowa+13 scores	.65	1.05	.58	1.94[c]	.54	2.10[c]	.82	3.23[d]
DAT (4 scores)	.59		.55		.29		.72	
DAT+13 scores	.64	0.46	.59	0.48	.55	2.07[c]	.85	3.36[d]

[a]Based on Guilford and Hoepfner (1971).
[b]Multiple *r*s not unbiased.
[c]Significant at the .05 level.
[d]Significant at the .01 level.

The data reported in Table 2.12 indicate that for two of the four courses, basic mathematics and noncollege algebra, information about scores on tests representing factors in the structure-of-intellect model is not particularly useful. However, increments in predictability are obtained for two courses, regular algebra and accelerated algebra, when tests representing factors are added to standardized academic ability composites.

The results reported in the Guilford, Hoepfner, and Peterson (1965) study must be accepted cautiously since the results are not cross-validated. Cross-validation refers to a kind of replication and is especially important in multivariate prediction studies. Strictly speaking, the Guilford *et al.* study is not a prediction study since no attempt was made to predict grades. Rather, grades were postdicted in that relationship between test scores and grades was discovered and defined after the grades were given. Such relationships must be accepted cautiously in a multivariate study because, given enough variables, by chance some of the variables are likely to discriminate among pupils with high and low grades. As the number of variables goes up, the increments achieved in predictability become increasingly suspect. Two procedures are available to overcome this problem and to establish rigorously the predictability inherent in a group of multivariate measures. In the first procedure, the subjects are randomly divided into two groups of approximately equal size. The first group is used to derive the weights to be assigned to test scores in order to maximize the value of the multiple correlation. The derived formula is tested in the second group of subjects from the same sample. The accuracy of the prediction in the second group is an indication of the amount of predictability to be obtained from these measures in this sample corrected for the opportunity to capitalize on chance variations. A second, more difficult, cross-validation is achieved by deriving a series of weights on tests from one sample and generalizing the results to a new and different sample. In this latter instance the generality of the multivariate prediction situation for different groups of subjects is tested.

Fortunately, we have two studies reporting on the attempt to cross-validate predictions of success in high school math from a knowledge of scores on tests representing factors in the structure-of-intellect model.

Holly and Michael (1972) used tests representing factors in the structure-of-intellect model to predict grades in high school algebra and performance on the Cooperative Math Test of competence in algebra. Their best composite of four tests from the structure-of-intellect model correlated with grades and Cooperative Math Test scores .57 and .64, respectively. These values when cross-validated by using the derived formula for the second half of subjects in the sample became .43 and .36, respectively. They report that the cross-validated correlations were about equal in value to the non-cross-validated correlations obtained when standard tests of mathematical aptitude were used to predict these criterion variables.

Caldwell *et all.* (1970) have performed a somewhat more rigorous cross-validation of the use of tests representing factors in the structure-of-intellect model to predict grades in high school geometry. They cross-validated their prediction formula derived from one high school sample on a second high school sample. They found that tests from the structure-of-intellect model could predict grades in high school geometry with a multiple correlation of .60 in both high schools. The addition of grades in high school algebra to the prediction equation led to increases in the multiple correlations to .65 and .70 in the two high schools. These are quite high values. However, on cross-validation in which the formula derived for one high school was applied to the second high school, there was a large drop in the values of the multiple correlations. The two multiple correlations of .60 dropped to values of .34 and .27, respectively. And, the multiple correlations including Grade 9 algebra scores dropped from .65 and .70 to .33 and .25, respectively.

The results of the studies by Caldwell *et al.* and Holly and Michael indicate that when cross-validated, the tests represented in the structure-of-intellect model do not achieve substantial predictability for grades. The values reported are of about the same order of magnitude as the predictability that would be achieved from scores on a standardized test of intelligence (see Lavin, 1965, for a review of research on the prediction of academic performance). The fact that scores on standardized tests of arithmetic achievement and grades in arithmetic courses are not predicted better by relatively factor-pure tests is at least compatible with the view that there is a limited number of rather general abilities that are represented by Guilford's tests. If this were so, it would explain why there is little increment in predictability obtained when more refined measures are added to the battery.

Conclusion

We have examined a number of different aspects of Guilford's structure-of-intellect model. And, on each issue we find little justification to assume that a structural model that postulates a great number of narrow specialized factors is correct. We found that, contrary to Guilford's assertions, his data, when critically examined, did not rule out a commitment to a single broad factor as a rough approximation of the structure-of-intellect. Second, we found that the factor structures reported in Guilford's research lacked replicability and clarity and suggested that the ability to sharply define and distinguish factors by tests with univocal substantial loadings was not present. Finally, we have seen that when cross-validated, tests representing the factors in the structure-of-intellect model are not more useful than more global measures of ability for the prediction of external criteria.

Conclusion

Our examination of the attempts to develop a more differentiated model for the structure of intellect has left us, in each instance, with the conclusion that the attempts have been less than successful. We found that Thurstone's research, contrary to his initial position, did not really contradict the legitimacy of Spearman's theory of general intelligence. Cattell's attempt to create a clear-cut distinction between g_f and g_c at the most abstract level led to a blurring of the distinction, with g_c placed in a subordinate role and the reaffirmation of a single intellectual ability. Finally, Guilford's attempt to develop a model for the structure of intellect without reliance on any construct that remotely resembled *g* was not empirically successful. Thus the available evidence clearly points in the direction of a single or pervasive ability factor as a major source of variance in measures of intellectual ability.

Notwithstanding the clarity of the evidence in favor of *g*, we are inclined to believe that a theory of the structure of intellect based on a single pervasive ability factor is an inadequate representation of the structure of intellect. We hold this view despite the consistency of the evidence in favor of *g* as a construct. The evidence we have reviewed is based on correlational analyses of the relationship among measures of ability. If each measure of ability that enters into a correlation matrix is a measure of the same ability (or abilities) in each individual who is given the measure, then the existing correlational data strongly imply that all tests of ability are measures of a single underlying ability. The assumption that each test or measure reflects the same underlying ability is usually an implicit assumption in correlation analysis (it is an explicit assumption in Thurstone's work). The factor-analytic approach to the understanding of the structure of intellect has always assumed that the meaning of a particular test can be understood in terms of its pattern of correlations with other tests. Tests that correlate presumably do so because they are measures of a common ability. The abilities measured by a test are assumed to be a property of the test and thus to be invariant across individuals. In view of this assumption, the positive correlations among all ability tests imply that all ability tests measure a common ability and that an individual's performance on any ability test is determined, in part, by his level of general ability.

Consider an alternative assumption. Assume that the same test can measure different abilities in different subjects. Consider a hypothetical example. Matrix algebra may be studied with the use of geometric analogues or it may be approached from the point of view of algebra, in which it is studied without respect to its geometric analogues. Success at learning matrix algebra might reflect skill in algebraic manipulations for some individuals and skill in geometric spatial abilities in others. If scores on a test of ability to learn matrix algebra are correlated with scores on tests related to ability to learn geometry and with scores on a test of skill at algebraic manipulation, then the correlations, for the sample as a whole, are likely to be positive. However, the positive correlations are misleading in that they are attributable to different subsets of the sample who are able to use the same abilities for two different tasks. Different subsets of individuals might account for positive relationships between different measures of ability by virtue of using the same abilities for different measures. This analysis is related to Guilford's conception of the nonunivocality of tests. So long as our subjects are ingenious enough to develop different methods and abilities to solve the problems psychologists present to them, it is likely that positive correlations between diverse tests will be obtained due to the discovery by subsets of individuals of algorithms and procedures that apply to different tests. This analysis suggests that the overwhelming statistical evidence for a single general ability factor does not imply that the structure of intellect is best conceived in terms of *g*. This suggests further that *g* should not be reified but rather that it should be thought of as a statistical abstraction. Finally, this analysis suggests that factor analysis may be a rather poor procedure for discovering the processes used by individuals to solve problems and, as such, it may not lead to an adequate understanding of the structure of human abilities.

3

Quantitative Characteristics of Intellectual Indices

This chapter will emphasize the following quantitative characteristics of intelligence test scores: the distribution of intelligence, relationship between intelligence scores obtained at two different times, and changes in scores related to age.

The Distribution of Ability

If people differ in their intelligence it should be possible theoretically to describe their differences in terms of a frequency distribution in which the frequency of occurrence of each amount of intelligence is noted (see Figure 3.1). What is the shape of this distribution? This is not an easy question to answer. In fact it may not have a unique answer. In order to understand why the question is difficult to answer it is necessary to consider a possible distinction between intelligence and that which is measured by tests of intelligence. Some psychologists have asserted that intelligence is that which intelligence tests measure. Different tests of intelligence tend to have somewhat different distributions. This implies that there are many distributions of intelligence and there is no way of determining which distribution is the correct one.

Other psychologists have tended to view intelligence as something different from that which is measured by any single test of intelligence. For example, those psychologists who accept some

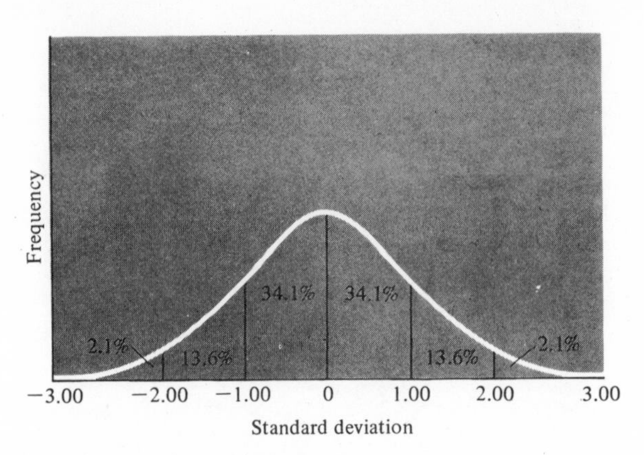

Figure 3.1. *Percentages of subjects in various regions of the normal distribution. (The percentages add up to 99.6 instead of 100 because a fraction of 1% of the subjects lie above and below 3 standard deviations.)*

version of the theory of *g* tend to view *g* as something theoretical that is different from what is measured by any one test or even by the set of existing tests. If this is so, the distribution of actual scores on a test (or tests) may not be the same as the distribution of intelligence.

It is important to emphasize the distinction between obtained and hypothetical distributions because test constructors have explicitly attempted to develop tests that conform to a preconceived arbitrary notion. It has been assumed, starting with Galton (1869), that the distribution of intelligence, like the distribution of such biological characteristics as height, is best described in terms of the familiar symmetrical bell-shaped normal curve (see Figure 3.1).

Research dealing with the distribution of the two most widely used individually administered tests of intelligence, the Wechsler and the Stanford-Binet, indicates that the distribution of scores on these tests is not normal. McNemar (1942) noted in the revision of the Stanford-Binet that the distribution that was obtained in the standardization sample departs from what would be expected on the basis of a normal curve. Burt gave the British version of the Stanfort-Binet to a representative sample of the British population. Figure 3.2 presents the distribution obtained by Burt (1963) and compares it to the normal curve. The distribution departs most dramatically from the normal with respect to the number of individuals at the "tails." There is an obvious excess of individuals with very low and very high scores. Since the frequency of occurrence of

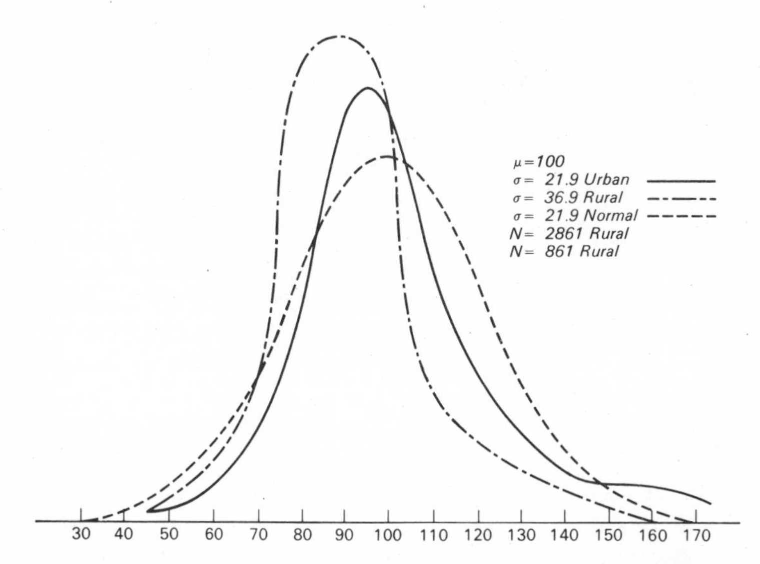

Figure 3.2. *Burt's distribution of intelligence. [From Burt (1963).]*

scores decreases with distance from the mean in a normal curve, the differences in the probability of occurrence of extreme scores between what would be expected on the assumption of normality and the obtained distribution are substantial. For example, very high IQs have a frequency of occurrence that is far more likely than would be expected on the basis of the assumption of normal distribution. It should also be noted that the distribution is not precisely symmetrical. There is a slight excess of scores at the low end.

Burt indicates that the obtained distribution can be described as a Pearson Type IV distribution. Wechsler (1944, p. 127) has also noted that the distribution of IQs on the Wechsler Adult Intelligence Scale can be described as a Pearson Type IV distribution. Burt explains the obtained distribution by appeal to a theory of genetic influences. The theory asserts that intelligence is dependent upon the combined effect of a large number of genes. If each of many genes have an equivalent effect, the resultant distribution would be normal. Burt next assumes that certain genes are likely to exert a larger influence than others. This effect would tend to produce a distribution with an excess, relative to the normal distribution, of individuals at the tails of the distribution. Third, one would expect that in as complex a characteristic as intelligence that any mutation or genetic error would tend to result in a decrement in intelligence. In addition, chromosomal irregularities, prenatal and birth damage all would combine to lower IQ. These factors would tend to produce a skewed curve with an excess of individuals in the lower part of the distribution. Cattell (1971) has also argued that the influence of assortative mating would tend to produce a distribution with a great number of individuals at the extremes. That is, individuals tend to marry individuals of equal intelligence and this would tend to increase the probability that favorable genes in one parent would co-occur with favorable genes in the other parent and that unfavorable genes would also co-occur with unfavorable genes leading to an increase of individuals in the next generation with genes that are either favorable or unfavorable for intelligence.

Burt's and Cattell's explanations of the distribution of intelligence are based on genetic influences. It should be noted that purely environmental explanations or an explanation involving the interaction of genetic and environmental influences can also be postulated which would help to explain the distributions. For example, one might assume that intelligence test scores result from the combined influence of a great variety of environmental events that might or might not be present. Some environmental influences might exert a larger influence than others. It is apparent that Burt's genetic model can be translated into an environmental model. We do not wish to prejudge our discussion of the influence of genetic and environmental effects on intelligence. We wish only to point out that the obtained distribution of IQ scores is compatible with a theory that emphasizes either genetic or environmental influences of intelligence test scores.

Finally, it should also be noted that there is likely to be a positive correlation between parental genetic and parental environmental influences. That is, parents with high genetic endowment for intelligence would be likely to provide an environment conducive to the development of high intelligence, and parents with low genetic endowment are likely to provide an environment inimical to the development of intellectual ability. The resulting hereditary–environmental correlation should also tend to produce a distribution that departs from the normal.

Although consistent data exist with respect to the distribution of intelligence on both the Stanford-Binet and Wechsler tests, one should be cautious in interpreting these obtained distributions as indicating the inherent distribution of the construct intelligence. Changes in either the environment or in gene pools from generation to generation would affect the distribution of intelligence. Also, it should be noted that intelligence tests cannot be completely divorced from a cultural context. Intelligence tests measure abilities germane to the acquisition of concepts and knowledge that are valued by a particular culture. As a culture changes and the definition of the knowledge that is valued within the culture changes, the composition of tests that are conceived of as measures of intellectual ability will change. And, as the content of tests changes, the distribution of ability defined by those tests may change.

It should also be noted that our intuitive concept, shaped by our cultural experience, of what constitutes an adequate distribution of ability is not unrelated to what is considered as a reasonable, empirically obtained distribution. And, our derived ability distributions shape our conception of what constitutes a reasonable social order. Our willingness to accept the kind of distribution that is obtained may be only a reflection of a culturally induced bias. It is possible to imagine social orders and requisite interdependent distributions radically different from those that have been obtained. For example, we might imagine a social order in which a small number of individuals require high intellectual ability—perhaps to program the culture—and all other members of the society engage in routine activities with minimal intellectual demand. Conversely, one could imagine a social order in which all complex functions are performed by machines and the only human functions are those that tend to require low intellectual ability. It is apparent that tests that would measure the underlying distribution of ability required for success in such imaginary cultures would have radically different distributions than those obtained from current tests.

Stability and Change in Intelligence

Intelligence test scores were originally age-corrected to form an intelligence quotient by dividing a score for mental ability by chronological age—MA/CA. Scores on contemporary tests of intelligence are treated differently. An individual's score is compared with the scores attained by other individuals of the same age. A person's position in the distribution of scores for his age is then converted into an IQ score based on his rank in the distribution. Such IQ scores are called *deviation IQs* and they have several advantages over the traditional IQ ratio. They are more appropriate for adult samples and they correct for possible changes in ability with age.

A good deal of research has dealt with the extent to which IQ scores are constant over a person's life. Using deviation IQs, such research focuses on the extent to which a person maintains or changes his position in the distribution of test scores at different ages. This question is typically answered by obtaining correlations between test scores obtained from the same individuals at two different ages.

The most systematic evidence related to this question comes from the Berkeley Growth Study. This is a longitudinal study of 61 children born between 1928 and 1929 which, as of this time, has continued through age 36. The sample was composed of "normal" children whose parents were white and English speaking, born in Berkeley, California, and is somewhat biased in that it is above average in socioeconomic status (see Jones and Bayley, 1941, for a description of the study group and sample). The study involved repeated testing of individuals from infancy through adulthood. Data with respect to the test–retest correlations of tests given at different ages with IQs at ages 17 and 18 are presented in Tables 3.1 and 3.2. A number of generalizations emerge from an examination of the data in Tables 3.1 and 3.2. IQs that are based on the average (mean) of three different testing occasions are better predictors of subsequent IQs than single IQ scores obtained at the same age period. This result is not unexpected. The average of three administrations permits one to correct for sources of nonrepeated error in a single

TABLE 3.1
Correlations between single test IQs at Different Ages and Age 18 in the Berkeley Growth Sample

Age	*r*
6	.77
7	.80
8	.85
9	.87
10	.86
11	.93
12	.89
13	.93
14	.89
15	.88
16	.94
17	.90

TABLE 3.2
Correlations between IQs Averaged over Different Ages and the Mean of IQs at 17 and 18 in the Berkeley Growth Sample

Average of months or years	*r*
Months	
1, 2, 3	.05
4, 5, 6	−.01
7, 8, 9	.20
10, 11, 12	.41
13, 14, 15	.23
18, 21, 24	.55
27, 30, 36	.54
42, 48, 54	.62
Years	
5, 6, 7	.86
8, 9, 10	.89
11, 12, 13	.96
14, 15, 16	.96

exam. Even when averaged over three occasions there is no relationship between tests given in the first 6 months of life and IQ at 18. There are very low positive correlations between IQs averaged over the 10- to 15-month period and IQ at age 18. It is not until the 18- to 24-month period that any appreciable degree of predictability can be achieved. What is perhaps most remarkable in these data is the relatively high degree of predictability that can be achieved by IQs given between ages 5 and 7. These IQs given at the very beginning of formal education predict IQ at the end of the high school period. Although the average of several tests does eliminate some of the unreliability in the measure, it does not eliminate unreliability completely. It is probably the case that the correlation of .86 between ages, 5, 6, and 7 and IQ achieved at 17 and 18, when corrected for attenuation, indicates that approximately 80% of the variance in IQs at age 17 and 18 is predictable at ages 5, 6, and 7. Also, the correlation of .96 between IQs given at ages 11, 12, and 13 and IQ at 17 and 18 indicates that, when corrected for attenuation, virtually all of the variance in the 17- and 18-year period is predictable from knowledge of IQs in the preadolescent period.

Although the degree of constancy is relatively high, a number of qualifications should be made. First, the results are based on a relatively limited sample and the relationships between IQ at early ages and later IQ are slightly higher than those obtained in other samples (see Bloom, 1964, for a discussion of these studies). Second, the sample is small and slightly biased. It would be difficult to generalize these results to nonwhite samples or to children who are low in socioeconomic status. Third, despite the high predictability achieved, individuals do change from administration to administration. Table 3.3 presents an indication of the amount of change obtained when comparing IQs obtained at different ages and IQ at age 17. Note that there is a consistent decrease in mean change as the age of testing moves closer to the comparison age of 17. Note further that single IQs—especially those given early in life—are quite capable of being substantially different from IQs obtained at a later age. Thus, in approximately 50% of the cases, IQs obtained at ages 6, 7, 8, or 9 will be 10 or more points different from IQs at age 17. And, over this period, the changes in the small sample of 40 can be as high as 25 or 30 points at any given age. Of course, if IQs averaged over three occasions were used, the amount of change for IQs given at different periods would be smaller.

TABLE 3.3
Changes in IQs Given at Different Ages and IQ at Age 17 in the Berkeley Growth Study

Age at testing	Range of changes	Mean change	Standard deviation of change
6 months	2–60	21.6	15.7
1 year	1–75	16.6	14.9
2 years	0–40	14.5	9.5
3 years	0–39	14.1	9.4
4 years	2–34	12.6	8.0
5 years	1–27	10.8	7.0
6 years	0–34	11.1	7.8
7 years	1–27	9.2	7.4
8 years	0–25	8.7	6.3
9 years	0–22	9.6	5.7
10 years	1–26	9.5	6.4
11 years	1–21	7.8	5.4
12 years	0–18	7.1	4.9
14 years	0–18	5.8	4.7

Anderson (Bloom, 1964) has developed an ingenious interpretation of the constancy of IQ obtained in the Berkeley Growth Study and in a similar study conducted at Harvard. The model used is essentially the same as one used to explain constancy in a characteristic such as height. One can assume that the amount of subsequent growth is unrelated to the amount of previous growth. A child at a particular age may be assumed to have reached a percentage of his final adult height. Even if the correlation between subsequent growth and previous growth is zero, there will be a positive correlation between height at one age and height at a subsequent age due to the initial advantage or disadvantage possessed by an individual at the previous age. That is, height at an earlier age overlaps height at a later age and may be thought of as the already attained growth plus a random (i.e., unrelated to previous growth) increment in growth. This model explains not only the correlation between previous height and subsequent height, but also the fact that the correlation increases as the difference decreases between the ages that are correlated. With respect to intelligence, the model implies that an increasing percentage of the total mental age attained by an individual is reached as the individual grows older. The level of mental age reached at any one age is unrelated to increments in mental age at subsequent ages. However, the overlap in mental age guarantees an increasing correlation between mental ages as the difference decreases in mental ages that are correlated.

Although Anderson's theory can be used to derive a close fit with the empirical results describing correlations between IQs given at different ages, there are a number of arguments suggesting that the overlap hypothesis is not a correct explanation of the constancy of IQ. On theoretical and on logical grounds it is unattractive to assume that intelligence is constant through the growth period. That is, the structure of intellectual abilities is probably different at different ages. An individual may develop new ways of attacking problems and new intellectual strategies as he becomes older. These lead to qualitative changes in intelligence as a function of age. Note that we do not measure intelligence at different ages with the same items. In this respect, intelligence is not like height, which can invariably be measured by the same procedure and invariably refers to the same thing.

An argument against Anderson's overlap hypothesis can be made on empirical grounds. The hypothesis assumes that increments in IQ are unrelated to previous intellectual level. Pinneau (1961) has analyzed the data from the Berkeley Growth Study and has concluded that individuals who are high in ability tend to show larger increments over a given unit of time than individuals who are low in ability (see Pinneau, 1961, especially Chap. 7).

Differential rates of gain in mental age are masked in conventional IQs and in correlations between them. If individuals of higher IQ show greater increments in ability relative to individuals who are low in IQ, there will be no change in their position relative to other individuals in the population. And such differential increments could occur even where the test–retest correlation in IQ is 1.00. In effect Pinneau's analysis suggests that individuals of high ability must have more growth in intellectual ability to maintain their relative position than individuals of low ability.

If increments in ability are larger for individuals high in ability than for individuals low in ability, we can explain these results by appeal to the constancy of influences that are likely to affect intelligence. The hereditary and environmental influences which combine to influence intelligence at earlier ages are quite likely to be present at later ages. Thus, most children who experience an environment that is favorable to intellectual growth early in life should continue to be in such an environment at later ages. This would explain why children who have made large amounts of progress would continue to do so.

Bloom (1964) has suggested that intellectual plasticity is greater early in life than later in life. He argues that later IQ is substantially determined by its overlap with the intellectual gains made at an earlier age. This suggests that interventions designed to increase IQ by providing a stimulating environment would be more effective when presented at an earlier, rather than a later, age. Bloom has suggested on the basis of the overlap hypothesis that there is a great deal of plasticity in intellectual ability up to age 4 but little thereafter. This implies that intervention programs designed to change intelligence should be confined to the first 4 years of life. By contrast, if the constancy of intellectual functioning is attributable to the constancy of influences that operate on the development of intelligence, it follows that there is no reason to believe that intellectual abilities are more plastic in one period rather than another. Thus the fact that the mean of the IQ scores at ages 17 and 18 is virtually perfectly predicted by the mean of the IQs at ages 11 to 13 does not imply on this view that intervention programs initiated at age 13 are doomed to failure. A plausible interpretation for the predictability between the age periods 11 to 13 and 17 to 18 would be that by age

11 to 13 the impacts of the school situation and of the home on the development of intellectual capacity have been experienced. There is little likelihood of a change in the school or home environment experienced up to age 13 with the environment that is experienced through ages 17 and 18. However, if interventions could be developed that changed the environment through this period or provided adolescents with experiences or skills missed at an earlier age, it is possible that large-scale changes in intellectual ability would result.

There is another argument that would suggest that early interventions to produce intellectual change would be more powerful than later interventions. If early intellectual ability is a precursor of the ability to develop further skills, then early interventions might have widespread effects that would influence subsequent learning. This view would imply that the effects of interventions to improve intellectual skills and abilities might have long-term effects and could only appropriately be assessed after a period of time had elapsed in which the newly acquired skill could operate and influence subsequent interactions with the environment.

The speculations we have indulged in with respect to the notion of plasticity in intelligence as a function of age may be seen as a substitute for the presence of actual data with respect to this question.

The Special Case of "Infant Intelligence"

Our examination of data from the Berkeley Growth Study indicates that there is relatively little relationship between scores on infant tests of intelligence and subsequent tests of intelligence. These and other analogous findings have led some psychologists to suggest that intelligence is not a fixed or invariant characteristic of an individual. Theorists such as Hunt (1961) would replace a model postulating invariance of intellectual ability with a model that assumes there are specific competencies whose attainments are preconditions for the development of subsequent, more advanced competencies. These competencies or skills may develop at different rates. Precocious development or attainment of some intellectual skill is not necessarily predictive of the ultimate level of attainment of an individual. This view, which is derived from a Piagetian conception of intellectual development, is also compatible with Guilford's theory in that it assumes relative independence among different intellectual abilities.

Such a view implies, first, that different measures of intellectual ability early in life will be unrelated to each other and, second, that scores on such measures will not be predictive of later intelligence. The data reviewed in Chapter 2 indicate that there is considerable evidence for the view that different measures of intellectual ability are substantially related to each other when given to school age children or adults. How do different measures of infant ability obtained during the first year of life relate to each other?

The data relating different measures of infant intelligence or ability are somewhat inconclusive for a number of reasons. There are relatively few studies that have been reported. Often the number of individuals included in the studies are relatively small. Most of the studies have correlated test results for two tests given at two different times. Since the performance of infants is likely to be variable on different days and times, the relationship between the two test scores may be influenced by this instability. The data taken as a whole do indicate that there is considerable overlap between measures of ability in the first 2 years of life. Thus, information obtained about intellectual ability from one test is likely to be at least positively (if not always substantially) related to other concurrently obtained indices of ability.

There are a number of studies that deal with relationships among different measures of ability in the first 2 years of life. McCall, Hogarty, and Hurlburt (1972) have reported the results of an item analysis at different ages of responses to the Gesell tests of infant development. They report there was a principal component or cluster of items that tended to be interrelated at each age tested. However, in no case did this principal component or cluster account for more than 19% of the total variance on the test. This finding suggests that omnibus tests of infant development tend to deal with somewhat more heterogeneous and unrelated abilities than tests appropriate for older children and adults. A comparable analysis at later ages would probably indicate that a principal cluster could be defined that would account for close to 50% of the common variance on the test.

In the last decade tests have appeared, based on Piaget's ideas about the development of intelligence, that can be used with preverbal children. Uzgiris and Hunt (1966) have developed such a test and Escalona and Corman (undated) have developed a test of the child's attainment of different levels of "object permanence." In this test, a child is shown an object that is hidden or displaced in several ways. The test measures the child's attainment of different stages of development culminating in the ability to be aware that an object that is covered and hidden is still present.

Several studies have been published relating scores on Piaget-type measures to other tests. Lewis and McGurk (1972) (see also Wilson, 1973, for criticism of this study) report correlations between the Bayley test and the Escalona-Corman test at 3, 6, 9, 12, 18, and 24 months of .24, .60, .16, .09, .23, and .02, respectively. Their sample was composed of 20 predominantly middle-class children. Golden and Birns (1968) report a correlation between the Cattell test of development and the Escalona-Corman test at 12 months of .24. King and Seegmiller (1973) have reported correlations between the Bayley mental scores and six scales derived from the Uzgiris-Hunt test at three age periods. Table 3.4 presents their results, and indicates that the correlations are quite variable. However, the bulk of the correlations are low and positive.

Gottfried and Brody (1975) have reported correlations among scores on the Bayley test of infant development and scores on the Escalona-Corman scale and scores on a measure of the development of schemas in relation to objects for a group of black children at 47 weeks of age. All tests were administered during the same session. Table 3.5 presents their results. Their data indicate that there is substantial agreement among these scores.

These data taken as a whole indicate that different measures of ability given during the first year of life are likely to be positively related to each other. There is at least some suggestion in the inconclusive literature reviewed here that the degree of relationship among these measures is somewhat lower than that which obtains at a later age. If this is so, it might be attributable to a greater independence of the development of intellectual skills or perhaps to the difficulties involved in testing preverbal children. In any case the data do not conclusively support a model of total independence of the rate of development of skills related to ability in the first 2 years of life.

The Berkeley Growth Study represents only one of several attempts to relate scores on infant tests of ability to scores obtained on tests given later in life. (For reviews of this literature see McCall, Hogarty, and Hurlburt, 1972; Rutter, 1970; and Stott and Ball, 1965.) The available data suggest that tests and measures other than the Bayley mental score may relate more substantially to later scores than would be indicated by the Bayley study. The available data do not permit one to assert with confidence that there is a lack of relationship between scores on tests of ability given during the first 18 months of life and later intelligence. There exist data that suggest there is substantial predictability for later intelligence test

TABLE 3.4

Correlations between Scores on the Bayley Mental Test and Scales from the Uzgiris-Hunt Test at Three Age Periods[a]

Age	Schemata in relation to objects	Schemata space	Object permanence	Causality	Means–ends	Initial imitation	Vocal imitation
14 Months	.32	.44	.04	.41	.42	.17	.03
18 Months	.37	—	−.12	.46	.05	−.33	.25
22 Months	.32	—	.14	.12	.36	—	.54

[a]Based on King, W.L., & Seegmiller, B. Performance of 14- to 22-month-old black, firstborn male infants on two tests of cognitive development: The Bayley Scales and The Infant Psychological Development Scale. *Developmental Psychology*, *3*, 323, 324. Copyright 1973 by The American Psychological Association. Adapted by permission.

TABLE 3.5
Correlations between the Bayley Scale of Mental Development and Two Piagetian-Type Scales[a]

	Bayley	Object permanence	Schemas
Bayley mental		.47	.84
Object permanence			.47
Schemata			

[a]From Gottfried, A.W., & Brody, N. Interrelationships between and correlates of psychometric and Piagetian scales of sensorimotor intelligence. *Developmental Psychology, 11,* 382. Copyright 1975 by The American Psychological Association. Reproduced by permission.

scores from scores obtained during the first year of life. Such data include evidence that some tests may be more predictive than the Bayley, and that certain measures based on parts of the Bayley or based on special experimental procedures may be predictive. In addition, few studies have attempted to predict later intelligence using a combination of scores on different measures obtained during infancy. Because of the uncertainty surrounding the issue of the relationship between early and late measures of intellectual ability, it is not possible to argue decisively that a measure of intellectual ability predictive of later intelligence cannot be obtained during the first year of life.

There is some evidence, although it is not consistent, that scores on the Gesell type of tests may be somewhat more predictive of later intelligence test scores than scores on the Bayley. Roberts and Sedgley (1966) report correlations between single administrations of the Griffiths tests given at 3 months, between 6 and 9 months, between 12 and 15 months, between 18 and 21 months, and IQs at age 7 of 0, .22, .39, and .49 for a sample of 54 normal children in England (see also Hindley, 1965). Data were not reported involving the averaging of several administrations of the Griffiths.

The Griffiths has not been used extensively in this country. It is, however, similar to the Gesell scales. (See Caldwell and Drachman, 1964, for evidence of substantial correlations between these measures.) Research relating the Gesell Developmental Schedule to later intellectual development has produced inconsistent results. Early research tended to report low correlations (Anderson, 1939). More recent research has tended to report higher correlations. Knobloch and Pasamanick (1960) have report higher correlations. Knobloch and Pasamanick (1960) have reported a correlation between scores on the Gesell tests administered at 40 weeks and scores on the Stanford-Binet test at age 3 of .48 for a sample of 195 children. McCall, Hogarty, and Hurlburt (1972) report correlations between the Gesell given at 6, 12, and 18 months and scores on the Stanford-Binet obtained at age 10 of .05, .37, .42, and .53, respectively, for girls and corresponding correlation of .07, .12, .27, and .59 for boys. These data indicate substantially lower predictive relationships for the Gesell (particularly for boys) than those obtained in the Knobloch and Pasamanick study. The data taken as a whole suggest that scores on the Gesell tests and the Griffiths tests, which are related to them, are somewhat more predictive of later intelligence than scores on the Bayley tests.

Although scores on infant tests considered as a whole are not substantially predictive of later intelligence test scores, it may be the case that subsets of items are predictive. The most dramatic evidence for such a possibility comes from a study involving the Berkeley Growth Study sample reported by Cameron, Livson, and Bayley (1967). They report that the age at which a child first passed items related to verbal skills during the 6-, 9-, and 12-month examinations was highly predictive of verbal IQ at age 26 for girls, $r = .74$. There was no relationship for boys. Similar results were obtained by Moore (1967) in England, He found that a verbal index derived from the Griffiths scale at 6, 12, and 18 months was predictive of subsequent IQ at 8 years for girls and not for boys. Thus, at least for girls, early verbal precocity seems predictive of latter intelligence. The greater predictability of females' later intelligence than of males' from scores obtained early in life has been occasionally noted (compare the results reported by McCall, Hogarty, and Hurlburt and see their article for a review of other relevant research).

Apart from the data reported by Cameron *et al.*, there are few available data indicating that subsets of items on standard tests given in infancy may achieve substantial predictability of later intelligence. However, it is possible that special procedures are necessary to measure cognitive ability in infancy. Lewis (1971) has reported a relationship between a measure of "response decrement" obtained at age 1 and 44-month Stanford-Binet IQ of .46 for girls and .50 for boys. In this connection, response decrement refers to a decrease in visual attention to a repeatedly presented pattern of lights. The decrement may be conceived of as an index of the rate at which an infant forms an internal representation or schema of the stimulus which is then used as a basis for matching a stimulus presentation with the internal representation or memory of the stimulus. When the internal representation or memory of the stimulus coincides or matches the percept elicited by the stimulus when it is presented, attention, as indexed by the time of visual fixation, is decreased. Thus, the rate of response decrement may be conceived of as a measure of the rate of information processing. (For reviews of relevant literature, see Lewis, 1971.)

The findings of Lewis (which require replication) combined with those of Cameron *et al.*, are at least suggestive of the possibility that adult IQ might be predicted on the basis of performance on measures obtained during the first year of life. Thus it appears premature to rule out the possibility of substantial continuity between preverbal intellectual ability and later ability.

These studies also suggest reasons for the lack of predictability between infant tests and those given later in life. Infant tests are designed for use with children who have not developed language. Later tests, even so-called nonverbal tests, assume that the person tested understands language. It may be the case that the attempt to assess intellectual ability in the preverbal child may require the more refined and complicated procedures used in laboratory settings rather than the use of items that can be administered without special equipment.

An additional reason for the lack of predictability of infant tests derives from the fact that performance on these tests is substantially influenced by physical and motor development. Gottfried and Brody have reported that measures of motor development, physical development, and activity level correlate substantially with performance on tests of ability at 48 weeks of age. However, measures of physical development, motor development, and activity level are correlated not at all or substantially lower with measures of ability obtained with school age children. Table 3.6 reports some of these correlations.

Motor development and scores on intelligence tests are unrelated in school age children (Dudek *et al.*, 1972; Singer, 1968). Gottfried and Brody report correlations at 48 weeks varying between .35 and .59. Height is weakly correlated with adult and school age intelligence with correlations ranging between 0 and .25 (Tanner, 1969). Gottfried and Brody report correlations ranging between .31 and .48. Activity level is either negatively related or not related at all to intelligence scores of school age children (Grinsted, 1939; Maccoby *et al.*, 1965). Gottfried and Brody report correlations ranging be-

TABLE 3.6
Correlations between Measures of Mental Development in Infancy and Other Infant Characteristics[a]

	Activity level	Length	Head circumference	Motor development (Bayley)	Mother's education	Father's education
Bayley Mental	.30	.48	.32	.58	.06	.03
Object permanence	.28	.31	.21	.35	.06	.00
Schemas	.32	.38	.40	.59	.03	−.02

[a]From Gottfried, A.W., & Brody, N. Interrelationships between and correlates of psychometric and Piagetian scales of sensorimotor intelligence. *Developmental Psychology, 11*, 383. Copyright 1975 by The American Psychological Association. Reproduced by permission.

tween .28 and .32. Finally, a number of investigators have reported correlations close to zero between measures of social class and tests of mental functioning in the first year of life (Golden & Birns, 1971). A large number of studies have reported significant correlations between social class background and intelligence test scores in school age and adult populations. These findings clearly indicate that the variables that relate to performance on infant tests are quite different from those that influence performance on tests given in later life. Furthermore, these data help explain why early and later measures are relatively unrelated.

Age-Related Changes in Intelligence

How do scores on intelligence tests change with age? Until recent years psychologists believed that scores on tests of intelligence declined during the adult years. The decline was assumed to begin in the third decade of life. Figure 3.3 is a representation of data obtained from Wechsler's 1958 standardization of the Wechsler Adult Intelligence Test. Figure 3.3 indicates a steady decline in intelligence test scores. This curve was used as a basis for the computation of deviation IQ scores. Because a deviation IQ represents the position of an individual in a distribution of individuals of the same age, it is clear that the same score on the WAIS test

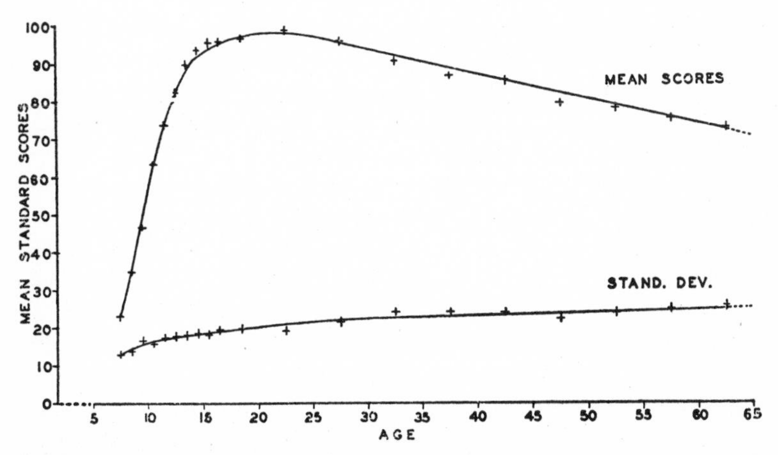

Figure 3.3. *Curve of mental growth and decline for Wechsler-Bellevue IQ scores. [From Wechsler (1944).]*

obtained at different ages will result in different IQs. Older individuals will receive higher IQs for the same raw score as younger individuals.

Recent research has tended to challenge the belief that performance on tests of intelligence declines with age. Most of the earlier studies used a "cross sectional" design. In this type of study individuals of different ages are given tests at approximately the same period of time. Time of testing is fixed. The ages of the persons tested varies. Differences between persons of different ages in this type of study involve comparisons between individuals who were born at different times. Any age-related changes obtained reflect both changes due to aging and changes due to differences in intelligence related to generational changes. In this sense the data are said to be confounded and cannot be used to infer the changes that are attributable to aging. In order to circumvent some of these difficulties "longitudinal" designs may be used. In this type of design the same individuals are tested at different times. Such a study is more difficult to do simply because of the duration of the research. Using a longitudinal design, comparison of intelligence at age 25 with intelligence at age 75 would take 50 years to complete.

Studies using longitudinal designs have consistently found little or no decrease in intelligence with age. A clear-cut example of the difference between the results obtained using cross sectional and longitudinal approaches can be seen in data obtained by Schaie and Strother (1968). Their study combined both approaches. They used a cross sectional procedure to test individuals ranging in age from 20 to 70 using the Thurstone Primary Mental Ability Test. Seven years later they reexamined 302 of the original 500 subjects in their sample, using the same test. Figure 3.4 presents their data. Note that in each case the cross sectional data show clear-cut declines with age. However, the longitudinal data indicate that there is a lack of decline. These data indicate quite clearly that intelligence test performance does not decline with age. Previous declines were evidently attributable to rather dramatic "cohort" or generational changes in ability. The exact reasons for these changes are not known. They may be due to an increase in the amount of education or perhaps to changes in nutrition and public health which influence performance on intelligence tests. In any case the data clearly demonstrate important environmental influences on intelligence test scores and clearly indicate that aging per se is not accompanied by a significant decline in the abilities measured by mental tests.

There are a number of other longitudinal studies that indicate a

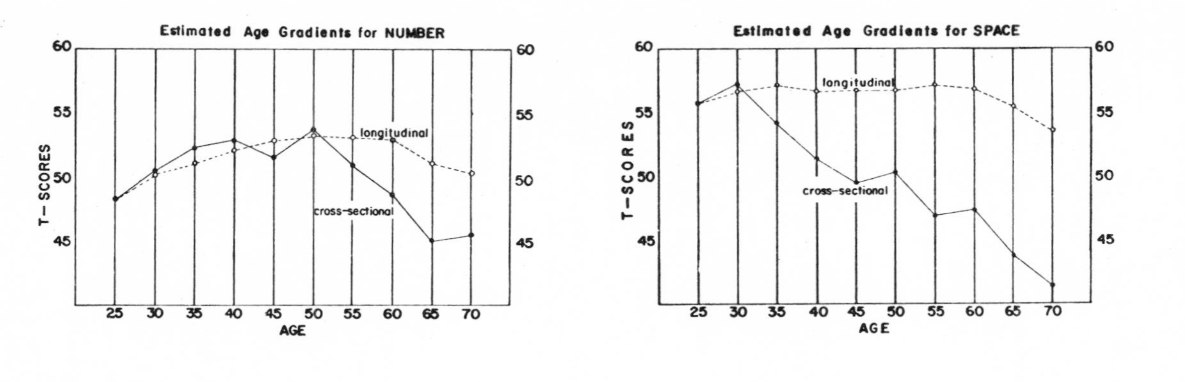

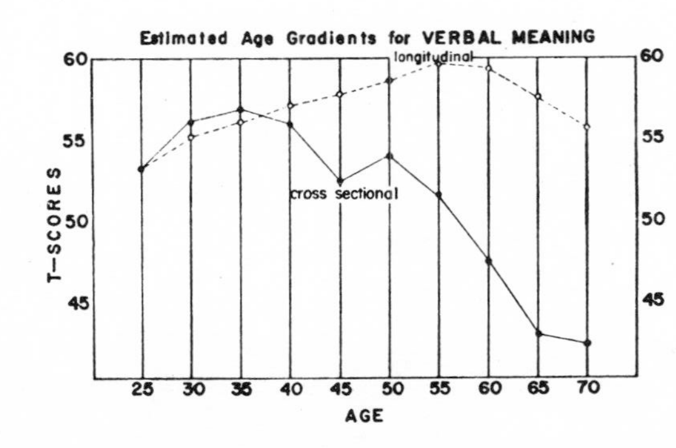

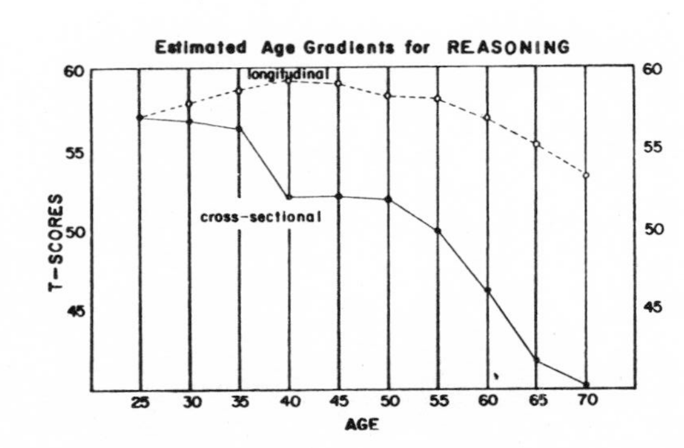

Figure 3.4. *Age gradients for intellectual ability. [From Schaie, K. W., and Strother, C. R. A cross sequential study of age changes in cognitive behavior.* **Psychological Bulletin,** 70, *675, 676. Copyright, 1968 by The American Psychological Association. Reproduced by permission.*

relative absence of decline in intelligence test performance with increasing age. Owens (1966) used the Army Alpha test to retest a group of males who were tested as college freshmen in 1919, in 1949 to 1950, and in 1961. He found that total scores on the test increased from 1919 to 1950 and then leveled off, showing only a slight decline from 1950 to 1961. There are some difficulties with the Owens study. Only 96 of a group of 127 subjects who were tested in 1950 were available in 1961. It is possible that the subjects who completed the study were not representative of the sample of subjects. Nevertheless, the data do clearly indicate that intelligence test scores do not inevitably decline with advancing age. (For other relevant longitudinal research, see Jarvik, Eisdorfer, and Blum, 1973; and Matarazzo, 1972, especially pp. 105–120.)

A number of investigators have attempted to extend longitudinal research on intellectual change into the seventh, eighth, and ninth decades of life. For example, Eisdorfer and Wilkie (1973) have reported the results of a 10-year longitudinal study of individuals whose age at the initiation of testing was between 60 and 79. For subjects who participated in the study through the 10-year period there were slight declines in scores on the Wechsler Adult Intelligence scale. (Note that the "survivors" of a 10-year study at this age are a nonrepresentative sample of those who began the study.) They analyzed changes in scores separately for individuals whose initial IQs were either low (L), middle (M), or high (H). Their data are presented in Figure 3.5 An examination of Figure 3.5 indicates that both groups (those in the 60 to 69 age group on initial testing and those in the 70 to 79 age group) showed some decline over the 10-year period. The older subjects showed the most decline. The amount of decline was not related to the initial level of ability. What is striking in these data is the relatively small magnitude of decline with increasing age. The average or mean decline for subjects in the 60 to 69 group who were between 70 and 79 on retest was less than four points and the decline of subjects in the 70 to 79 age group who were between 80 and 89 on retesting averaged less than eight points. These data indicate that there are declines in intelligence test score performance in the eighth and ninth decades of life, but they are of a relatively modest magnitude. (For a review of other relevant research see Jarvik, Eisdorfer, and Blum, 1973.)

It is now generally accepted that there is little decline in scores on intelligence tests through the life span until advanced age when there are small declines. More recent research has focused on two more refined issues. First, what characteristics of individuals predict

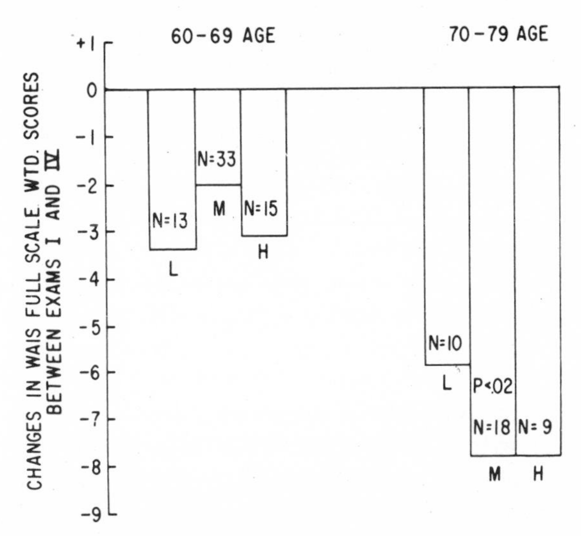

Figure 3.5. *Longitudinal change (delta scores) over a 10-year period as measured by the Wechsler Adult Intelligence Scale (WAIS) among individuals initially examined at ages 60-69 and 70-79, with either low, middle, or high WAIS scores on the initial examination. [From Eisdorfer and Wilkie (1973).]*

their eventual decline or increase in intelligence? Second, are changes in intelligence test performance the same for all types of tests? Or, is it the case that age-related changes in performance are different for different types of tests? Through the adult years the longitudinal research shows that there is relatively little relationship between initial level of intelligence and the direction of change in intelligence (see the data in Figure 3.5). However, there is some evidence that the kind of education received during the school age years is related to changes in performance on intelligence tests. Härnqvist (1968a,b) has reported an extremely sophisticated study of this problem. He used data representing a 10% sample of all of the male pupils in Swedish schools who were born in 1948. These pupils were tested when they were 13 and again at 18 as part of their military examinations. Using a rather complex statistical procedure (canonical correlations to obtain components on tests with different subscales, and a procedure comparing regression weights on these components), he was able to show that the subjects who received more extensive and more academic education tended to show greater positive changes in performance than did the subjects whose educational experiences were more limited and less academically oriented. His estimate of the difference in change scores between pupils whose educational experience did not exceed what was compulsory in Sweden and pupils whose educational experiences were at the upper level, including some gymnasium experience, was about 62% of the standard deviation of IQ scores or about 10 IQ points. These data suggest that initial differences in scores on tests that are related to the selection of different educational experiences are enlarged as a result of these educational experiences.

Härnqvist went on to relate changes in intelligence between 13 and 18 to differences in family background. He found that the influence of family background on changes in intellectual functioning was smaller than the influence of education.

The results of Härnqvist's study are clear-cut. In addition, his study is based on a large representative sample and his data were analyzed in a sophisticated manner. However, his conclusions may in part be limited to a school system in which there is considerable tracking and segregation of students. Good data indicating the influence of educational experiences on changes in intelligence test scores in the United States are not readily available (see Jencks, 1972, especially Chap. 3).

However, there are considerable variations among academic experiences of high school students in the United States (e.g., students pursuing academic as opposed to vocational education) and it is not unreasonable to assume that the kind of education obtained during the school years influences changes in intelligence test scores.

Changes in intelligence test scores among the aged have been related to physical changes. The most dramatic evidence of this comes from studies showing that declines in intelligence test scores are predictive of imminent death. For example, Reimanis and Green (1971) obtained intelligence test scores at age 68 and compared them with those obtained 5 to 10 years earlier by a group of hospitalized veterans. They found that those individuals who died within the next 12 months had suffered a decline of 15.50 points in IQ during the previous 5 to 10 years. Those individuals who survived more than 2 years had an average decline of 5.67 points during the5- to 10-year period preceding their retest at age 68.

These findings are related to other findings that have attempted to relate changes in intellectual functioning among the aged to physical conditions. Wilkie and Eisdorfer (1973) obtained diastolic blood pressure for subjects in their 10-year longitudinal study of intellectual functioning in the aged. They found that declines in intellectual functioning for subjects who were between ages 60 and

69 occurred only in the group of 10 subjects who had high blood pressure.

Wang (1973) found a correlation of .43 between changes in the performance (or nonverbal part of the Wechsler tests) over a 3-year period and a measure obtained from the EEG of the mean frequency of dominant rhythms for a group of individuals whose average age was 70.1 on first testing.

In another part of this study Wang obtained a measure of the amount of cortical blood flow. He found, using a group of subjects with a mean age of 79.5, that the decline in full scale IQ over the previous 12 years was 1.8 for the 12 subjects in his sample with relatively high cortical blood flow and 12.3 points for the 12 subjects with relatively low cortical blood flow. These data suggest that intellectual declines among the aged are not a general accompaniment of aging but rather depend on the physical condition of the aged person. Age is itself not the critical variable. Age is, however, related, although not perfectly, to physical changes. And the extent and kind of changes that occur evidently relate to ability to survive and also to changes in performance on tests.

In addition to the question of differential changes among different individuals, investigators have been concerned with the question of differential changes in types of tests with age. Cattell's g_f–g_c theory provides a theoretical structure for the examination of this question. Cattell's theory asserts that g_c reflects the process of acculturation. There is no reason to assume that g_c will decline with age; indeed, it may increase. On the other hand, g_f is dependent on the biological state of the organism. It is assumed to decline starting with the end of the second decade in life. Horn (1970) has attempted to organize the available data on changes in intelligence test performance with age in terms of the g_f–g_c theory. He points out that tests such as vocabulary, that clearly reflect g_c show the least decline with age in cross sectional research. Performance-oriented tests, some of which might reflect g_f, tend to show more decline with age in cross sectional studies. Horn and Cattell (1967) have performed one of the best cross sectional studies dealing with this problem. They defined a g_f and a g_c score for a set of primary abilities tests (French, 1963). The cross sectional changes in their composite g_f and g_c scores are presented in Figure 3.6. Figure 3.6 indicates a clear decline in g_f scores from the teenage period through late adulthood. Composite scores for g_c are increased during these years. The omnibus or total intelligence, composed of the combined score for both components, remains relatively unchanged.

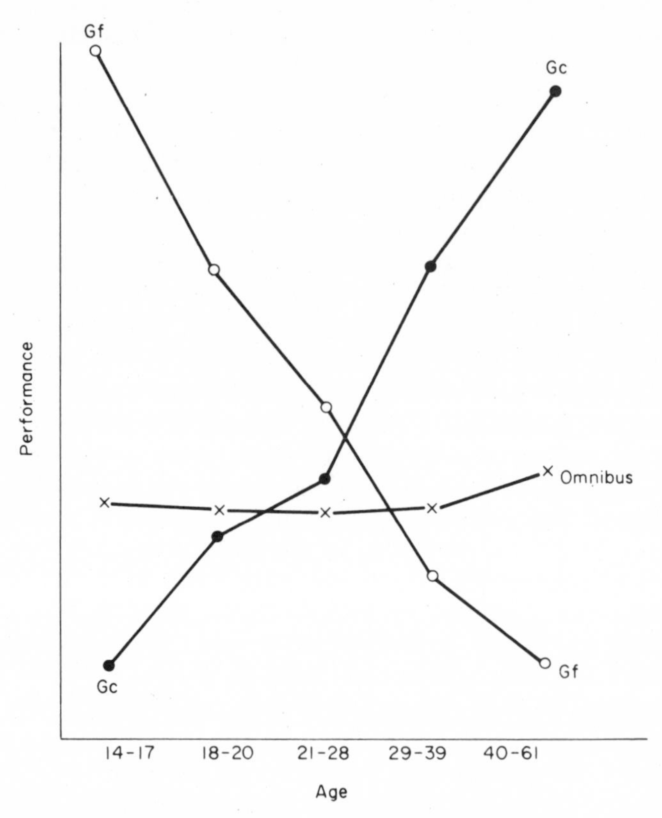

Figure 3.6. *Performance as a function of age. Summary of covariance analysis reults. Sex, education, visualization, and fluency were statistically controlled. [Based on Horn and Cattell (1967).]*

The data in Figure 3.6 present clear-cut evidence in favor of the separation of g_f and g_c. The data must be accepted cautiously. First, the review of research on the g_f–g_c distinction presented in Chapter 2 indicates that there is some ambiguity in defining which tests load consistently on one or the other factor. More importantly, the results reported by Horn and Cattell represent a cross sectional analysis. There is little direct evidence using tests that are presumed to be clear measures of g_f (e.g., the Cattell culture-fair tests) in a longitudinal study that g_f declines after age 20. Cattell (1970) cites an unpublished study by Wackwitz, based on an analysis of the Schaie and Strother (1968) data and the Horn and Cattell (1967) data, which indicates that, with control for generational changes, there may be little or no decline in g_f with age. Schaie (1970) has also attempted to relate age changes in ability to the g_f–g_c distinction. Figure 3.7 presents his analysis of data involving an estimate in longitudinal changes on four types of tests of ability. His data indicate sharp expected declines in measures of psychomotor speed and word fluency and lack of decline in measures of spatial ability and verbal reasoning. He suggests that the declines in the measure of speed and word fluency may be related to the fact that they are related to g_f. However, there is little evidence that measures of

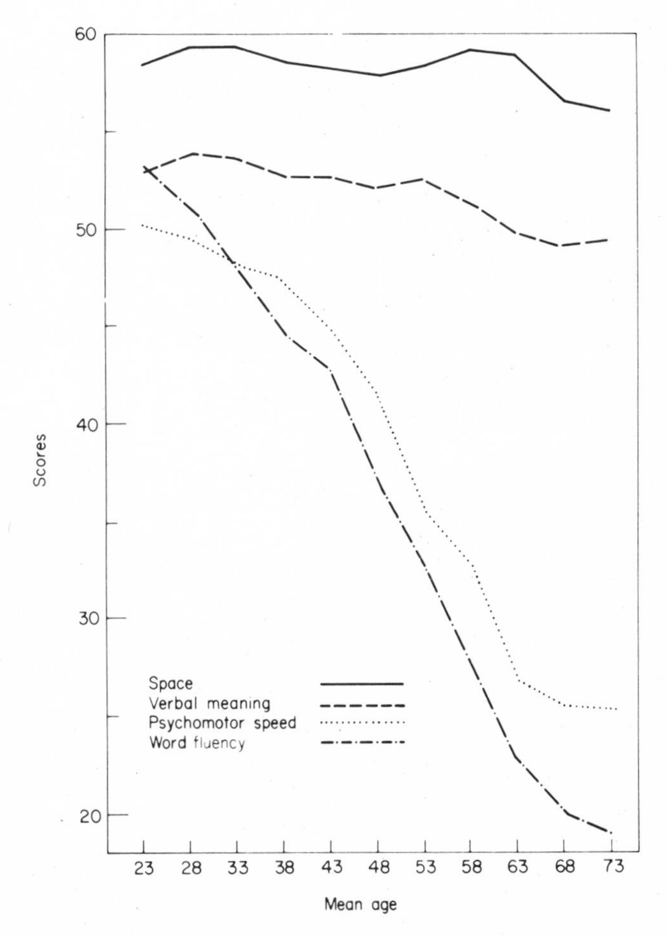

Figure 3.7. *Estimated longitudinal Gradients. Cohort birth year, 1940. [From Schaie (1970).]*

psychomotor speed and word fluency are clear indicators of g_f (see the data reviewed in Chapter 2). In fact, spatial ability is sometimes, although not invariably, related to g_f.

In summary, there is little or no evidence in longitudinal research indicating clear-cut declines in unambiguous measures of g_f as a function of age.

4

Intelligence and Achievement

Our discussion of scores on intelligence tests has circumvented a critical question. Do intelligence tests measure anything useful? We have examined relationships among intelligence tests, we have examined changes in scores as a function of age, and we have discussed the distribution of scores on such tests. None of this material provides direct information about the validity of the tests.

The use of the term "intelligence" to describe what is measured by intelligence tests biases our answer to the question of the usefulness of intelligence tests, since we are inclined to believe that intelligence is an important characteristic of a person. However, the mere fact that tests of intelligence are given that name does not indicate that the name is deserved or well chosen. There is no simple clear-cut procedure that permits one to determine that tests of intelligence are in fact measures of intelligence. Note that this assertion presupposes that intelligence—however defined—is a hypothetical characteristic of a person which transcends his score on intelligence tests. Such scores may or may not accurately index the degree of the hypothetical variable, intelligence, possessed by a person.

One way of determining what is measured by tests of intelligence is to examine the relationship between scores on intelligence tests and other measures. The total set of relationships into which such scores enter would then eventually serve to increase our understanding of what is measured by such tests. After examination of this network of laws and relations we would assign to the hypothetical variable just those properties required to explain the obtained relationships.

For convenience in exposition we can distinguish between variables that are conceived of as being consequences of intelligence, i.e., those things that are influenced by intelligence and those variables conceived of as determinants—i.e., as variables that influence intelligence. The distinction is arbitrary. For example, success in school is usually conceived of as a consequence of intelligence. However, data exist that suggest success in school acts as a determinant of intelligence in that children assigned to educational tracks for academically talented children show somewhat greater gains in intelligence than children not assigned to such tracks.

Despite the ambiguities inherent in the direction of influence or causality, we will deal primarily in this chapter with what may be conceived of as consequences of intelligence test scores.

Intelligence and School Success

The earliest evidence for the predictive validity of intelligence tests scores derives from studies showing a relationship between such scores and measures of school success—either grades or scores on tests purporting to indicate what has been learned in the schools. The relationship has been consistently confirmed. (For a review of studies on this issue, see Lavin, 1965.) The correlation between scores on intelligence tests (typically group tests) and grades is about .50. The correlations reported usually tend to be somewhat higher in the elementary school years and in high school than in college. For example, Hinkelman (1955) reports correlations of .65 between grades and intelligence in grades 2 to 7. The lower value of the correlations typically reported for the relationship between IQ and grades in college is probably attributable to the restriction of range of talent with respect to test scores found in college populations.

The correlation between scores on intelligence tests and school performance tells us very little about the meaning of what is measured by tests. The interpretation of the meaning of this relationship is fraught with difficulty. Furthermore, since public schools for the most part are not selective institutions and are forced to accept virtually all children residing in a particular geographic area, the obtained correlation is of little use for the purpose of selecting students. And, the relationship is of little use to the schools since they have information about grades. A measure that predicts grades with a correlation of .6 is not a useful surrogate for grades. Therefore, the fact that intelligence tests predict school grades is of little practical or theoretical interest.[1]

From a theoretical point of view, ever since Binet we have tended to think of intelligence tests as measures of something distinct from school achievement. One way of drawing the distinction between what is purportedly measured by intelligence tests and school achievement is in terms of the distinction between aptitude or ability and achievement. The latter reflects what is learned in the schools. The former reflects an ability or capacity to learn which may or may not be used by an individual. Presumably ability or aptitude is an amalgam of genetic endowment and previous learning experiences which determine the total set of intellectual skills and capacities possessed by an individual relating to his ability to acquire specific knowledge taught in the schools. The distinction between ability or aptitude and achievement is not categorical, but is best conceived of as falling on a continuum. Every measure of ability or aptitude must reflect past learning and achievement. Different measures may reflect different degrees of mixtures between ability and achievements (compare Cattell's distinction between g_f and g_c).

If intelligence tests are theoretically assumed to be measures of something different from school achievement, then the discovery of a correlation relating test scores to achievement does not establish that the tests measure something different. One way of empirically distinguishing between what is measured by tests of intelligence as opposed to measures of achievement is with reference to an assumed time lag in the direction of influence of these variables. Ability to learn is undoubtedly influenced by past learning achievements. And such ability at a particular time may be assumed to be a determinant of future learning experiences which in turn may influence future ability to learn. This analysis suggests that there is a time lag in the direction of causal influence of these variables. Ability at Time 1 can influence subsequent achievement at Time 2, which in turn can influence later ability at Time 3, etc., where t_3 is after t_2 which is after t_1. However, achievement at t_1 cannot directly influence ability at t_1. Thus, ability must influence achievement before achievement can influence ability. This "time lag" hypothesis can be tested by cross-lag panel analysis. Crano, Kenny, and Campbell (1972) ob-

[1] The relationship is of practical importance at the college level where tests of ability may be used to select students with high probability of getting good grades.

tained correlations between measures of achievement and measures of ability in a large sample of suburban school children attending Milwaukee public schools. They found that the Thorndike-Lorge IQ test at Grade 4 predicted a composite index of tests of school achievement in Grade 6 with a correlation of .73. The correlation in this sample between a comparable index of achievement in Grade 4 and intelligence test scores in Grade 6 was .70. The differences in the magnitude of correlation were significant although they were not large. The difference in the correlations indicates that a measure of ability was in fact capable of predicting subsequent achievement better than a measure of achievement was capable of predicting subsequent ability. This finding is compatible with the view that tests of intelligence do in fact measure something distinct from school learning.

Scores on intelligence tests have been related to subsequent achievement in schools. Benson (1942) has reported data for 1989 pupils in the sixth grade in Minneapolis. She obtained data on the highest grade completed for 1680 of these pupils and found a correlation of .57 between IQ of Grade 6 and number of years of education. (For a review of other relevant studies, see Tyler, 1956.) Correlations between intelligence test scores and number of years of education completed for adult sample are approximately .70. (See Miner, 1957; Wechsler, 1958, who reported correlations of .69, .66, and .72 for groups of 18- to 19-year-olds, 25- to 34-year-olds, and 45- to 54-year-olds, respectively, in his WAIS standardization sample.) These data clearly indicate that individuals who score high on tests of intelligence are more likely to continue their education than individuals who score low on such tests.

The correlation between intelligence test score and number of years of education completed is, in part, dependent upon the fact that both variables are positively related to parental education and socioeconomic level. Parents who have completed many years of education and whose socioeconomic status is high are likely to have children who score high on tests of intelligence and who are likely to continue their education beyond minimal levels. However, the relationship is not solely dependent upon parental encouragement to continue. Students with low IQ who attempt to continue their education are less likely to be successful than students with high IQs. For example, Hartson and Sprow (1941) found that 65% of freshmen whose IQ was less than 110 failed to graduate from Oberlin College. Marshall (1943) reported comparable data for Franklin and Marshall College. Such results do indicate, however, that a substantial number of individuals with IQs less than 110 are able to complete and graduate from colleges of high academic standing. This proportion (approximately 35%) would probably be higher for colleges with somewhat lower academic standards and for colleges with special provisions for assisting students whose academic aptitude is apparently somewhat below average. However, the relationship between IQ and probability of completion of college does help to explain the substantial correlation between intelligence test scores and number of years of education in adult samples.

Intelligence and Occupational Status

We have seen that scores on intelligence tests relate to number of years of education completed. We know also that the number of years of education completed is an important determinant of occupational status in American society. Accordingly, we should expect that there would be some relationship between intelligence test score and occupational status. This relationship has been dealt with in a number of ways. Duncan, Featherman, and Duncan (1972) have dealt with this relationship. Their study began with an analysis of data collected by Barr (see Terman, 1925). Barr asked 30 judges to rate the intelligence required for success at a variety of occupations. Each occupation was assigned a score based on the composite judgments of the 30 judges. These scores were correlated with ratings of socioeconomic prestige or status associated with these occupations. Depending on the measure of status used, the correlation between these ratings is either .81 or .90. These findings indicate that ratings of intelligence required to succeed at an occupation are highly correlated with the socioeconomic status or prestige associated with the occupation. Commenting on this finding, Duncan, Featherman, and Duncan assert:

> The reconstruction we wish to suggest is the following. Every society implicitly designates certain key roles in which performance is variable, with the quality of the performance being a basis for the assignment of status. (Other statuses, of course, may depend upon factors besides performance—the so-called ascribed statuses.) Where the society is one with a complex division of labor, many differentiated occupations are pursued, and these occupations are highly salient among the key roles whose pursuit is a basis for status achievement. Adequate performance in a high status occupation is taken by the social group as prima facie evidence of social capability. However, poor

> performance in a high status occupation leads to uncertain tenure of the status, and performance—whether good, bad, or indifferent—of a low status occupational role is not seen as providing any sizable increment to consensual estimates of a person's value to society. What we call "occupational prestige" corresponds to an unmistakable social fact. When psychologists came to propose operational counterparts to the notion of intelligence, or to devise measures thereof, they wittingly or unwittingly looked for indicators of capability to function in the system of key roles in the society. What they took to be mental performance might equally well have been described as role performance. Indeed, it was clear in the minds of the pioneers of mental testing that they wished to tap capacity to perform well in another social situation—that of the school. For their immediate purposes, it was unnecessary to expand upon the sociological observation that the school is itself (among other things) a primary mechanism for selecting incumbents of occupational roles.
>
> Our argument tends to imply that a correlation between IQ and occupational achievement was more or less built into IQ tests, by virtue of the psychologists' implicit acceptance of the social standards of the general populace. Had the first IQ tests been devised in a hunting culture, "general intelligence" might well have turned out to involve visual acuity and running speed rather than vocabulary and symbol manipulation. As it was, the concept of intelligence arose in a society where high status accrued to occupations involving the latter in large measure so that what we now *mean* by intelligence is something like the probability of acceptable performance (given the opportunity) in occupations varying in social status [pp. 78–79].

Not only is there a positive relationship between ratings of prestige and intelligence required for certain occupations, there is a strong relationship between the obtained average level of intelligence of members of an occupation and the social prestige of the occupation. Table 4.1 presents data collected by Harrell and Harrell (1945) based on Army General Classification Tests during World War II. The median intelligence test score of recruits engaged in various occupations clearly increases in a roughly monotonic fashion with increases in the social prestige assigned to various occupations. One other aspect of these data is striking. The variability of the scores increases as one goes down the list. There is relatively little difference in the upper level scores for different occupational groups. However, as the average or median intelligence found in an occupation increases, the lower limit of intelligence found in the occupation increases. Since the upper level remains relatively constant the result is a decrease in the variance of scores of individuals engaged in that occupation. These data fit a model suggesting that intelligence acts as a variable that sets a threshold for occupational entry. Our social system

TABLE 4.1

Mean AGCT Standard Scores (IQ), Standard Deviations, and Range of Scores on 18,782 Army Air Force White Enlisted Men by Civilian Occupation[a]

Occupation	N	M	Median	Standard deviation	Range
Accountant	172	128.1	128.1	11.7	94–157
Lawyer	94	127.6	126.8	10.9	96–157
Engineer	39	126.6	125.8	11.7	100–151
Public relations man	42	126.0	125.5	11.4	100–149
Auditor	62	125.9	125.5	11.2	98–151
Chemist	21	124.8	124.5	13.8	102–153
Reporter	45	124.5	125.7	11.7	100–157
Chief clerk	165	124.2	124.5	11.7	88–153
Teacher	256	122.8	123.7	12.8	76–155
Draftsman	153	122.0	121.7	12.8	74–155
Stenographer	147	121.0	121.4	12.5	66–151
Pharmacist	58	120.5	124.0	15.2	76–149
Tabulating machine operator	140	120.1	119.8	13.3	80–151
Bookkeeper	272	120.0	119.7	13.1	70–157
Manager, sales	42	119.0	120.7	11.5	90–137
Purchasing agent	98	118.7	119.2	12.9	82–153
Manager, production	34	118.1	117.0	16.0	82–153
Photographer	95	117.6	119.8	13.9	66–147
Clerk, general	496	117.5	117.9	13.0	68–155
Clerk-typist	468	116.8	117.3	12.0	80–147
Manager, miscellaneous	235	116.0	117.5	14.8	60–151
Installer-repairman, tel. & tel.	96	115.8	116.3	13.1	76–149
Cashier	111	115.8	116.8	11.9	80–145
Instrument repairman	47	115.5	115.8	11.9	82–141
Radio repairman	267	115.3	116.5	14.5	56–151
Printer, job pressman, lithographic pressman	132	115.1	116.7	14.3	60–149
Salesman	494	115.1	116.2	15.7	60–153
Artist	48	114.9	115.4	11.2	82–139
Manager, retail store	420	114.0	116.2	15.7	52–151
Laboratory assistant	128	113.4	114.0	14.6	76–147
Took-maker	60	112.5	111.6	12.5	76–143
Inspector	358	112.3	113.1	15.7	54–147
Stock clerk	490	111.8	113.0	16.3	54–151
Receiving and shipping clerk	486	111.3	113.4	16.4	58–155
Musician	157	110.9	112.8	15.9	56–147

Continued

TABLE 4.1—*Continued*

Occupation	N	M	Median	Standard deviation	Range
Machinist	456	110.1	110.8	16.1	38–153
Foreman	298	109.8	111.4	16.7	60–151
Watchmaker	56	109.8	113.0	14.7	68–147
Airplane mechanic	235	109.3	110.5	14.9	66–147
Sales clerk	492	109.2	110.4	16.3	42–149
Electrician	289	109.0	110.6	15.2	64–149
Lathe operator	172	108.5	109.4	15.5	64–147
Receiving and shipping checker	281	107.6	108.9	15.8	52–151
Sheet metal worker	498	107.5	108.1	15.3	62–153
Lineman, power and tel. & tel.	77	107.1	108.8	15.5	70–133
Assembler	498	106.3	106.6	14.6	48–145
Mechanic	421	106.3	108.3	16.0	60–155
Machine operator	486	104.8	105.7	17.1	42–151
Auto serviceman	539	104.2	105.9	16.7	30–141
Riveter	239	104.1	105.3	15.1	50–141
Cabinetmaker	48	103.5	104.7	15.9	66–127
Upholsterer	59	103.3	105.8	14.5	68–131
Butcher	259	102.9	104.8	17.1	42–147
Plumber	128	102.7	104.8	16.0	56–139
Bartender	98	102.2	105.0	16.6	56–137
Carpenter, construction	451	102.1	104.1	19.5	42–147
Pipe fitter	72	101.9	105.2	18.0	56–139
Welder	493	101.8	103.7	16.1	48–147
Auto mechanic	466	101.3	101.8	17.0	48–151
Molder	79	101.1	105.5	20.2	48–137
Chauffeur	194	100.8	103.0	18.4	46–143
Tractor driver	354	99.5	101.6	19.1	42–147
Painter, general	440	98.3	100.1	18.7	38–147
Crane-hoist operator	99	97.9	99.1	16.6	58–147
Cook and baker	436	97.2	99.5	20.8	20–147
Weaver	56	97.0	97.3	17.7	50–135
Truck driver	817	96.2	97.8	19.7	16–149
Laborer	856	95.8	97.7	20.1	26–145
Barber	103	95.3	98.1	20.5	42–141
Lumberjack	59	94.7	96.5	19.8	46–137
Farmer	700	92.7	93.4	21.8	24–147
Farmhand	817	91.4	94.0	20.7	24–141
Miner	156	90.6	92.0	20.1	42–139
Teamster	77	87.7	89.0	19.6	46–145

[a] Based on Harrell and Harrell (1945).

apparently acts in such a way that individuals who score low on intelligence tests have limited opportunity to enter more prestigeful occupations.

The results reported in Table 4.1 are further buttressed by the findings of Ball (1938), who did a longitudinal study relating intelligence test scores to subsequent occupational achievement. Ball obtained intelligence test scores for young men in 1918 or 1923 and then related these scores to occupational data obtained in 1937. He found correlations between ratings of occupational prestige and intelligence test score of .71 and .57 for the 1918 and 1923 samples, respectively. These data indicate that intelligence test scores are good predictors of ability to enter prestige occupations.

Since occupational prestige and income are associated, one would expect that intelligence test scores and income would be associated. Relatively little research has been done directly dealing with this question. Jencks (1972, see especially Appendix B) has summarized the available data and concluded that the correlation between intelligence test score in early adulthood and adult income is approximately .31.

Our discussion of the relationship between intelligence test scores and educational attainments, occupational status, and income has considered each of these as a separate and independent relationship. However, all of these variable are interrelated. All are, for example, influenced by parental education and occupation. In recent years attempts have been made to consider simultaneously within one conceptual framework or model all of the interrelationships that exist among these variables. The method used for this purpose has been *path analysis* (see Blalock, 1964, and Duncan, 1966, for an exposition of the method). The statistical procedures involved in the method are somewhat complex. The end results, however, are not difficult to grasp. A path analysis involves assumptions about the direction of causal influence among a set of variables. The end result of the analysis is dependent upon these assumptions. The resultant model attempts to indicate the extent to which a particular variable influences another when all of the other variables under consideration are held constant or are statistically controlled. Furthermore, the model provides estimates of the extent to which a variable exerts a direct and an indirect influence on some dependent variable. Duncan, Featherman, and Duncan have developed path analyses linking intelligence and occupational status. Figure 4.1 presents one of their models. The variables to the left of the diagrams formed by double-arrowed curved lines are taken as independent variables. Rela-

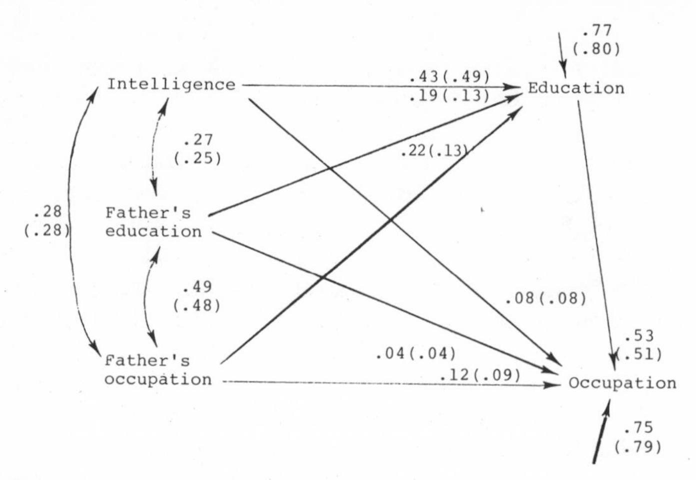

Figure 4.1 *Abridged version of the final model of ability and achievement with path coefficients estimated for two populations. [Based on Duncan, Featherman, and Duncan (1972).]*

tions among these variables are not analyzed—i.e., they are taken as given. The numbers adjacent to the curved lines represent the values for simple correlations between the variables. (The numbers in parentheses represent values derived from a second, independent data source.)

Thus the correlation between intelligence and father's education in these data is .27. Each of the "independent" variables are considered to be determinants of, or equivalently, causally linked, to education and ultimately to occupational status. Straight lines with single arrows indicate the direction of influence assumed in the model. The values adjacent to such straight lines represent "path coefficients" and these values may be taken to represent the influence of one variable upon another when all other relevant variables present are controlled. Figure 4.1 indicates that intelligence score during adolescence is a more potent influence on ultimate educational level than parents' educational level or occupational status (.43 versus .19 and .22 respectively). Education, in turn, is the single most important determinant of occupational status and has a path coefficient for this variable of .53. Furthermore, the model indicates that each of the independent variables primarily exerts its influence on occupational status indirectly by way of their influence on educational attainment. Thus, the direct path between intelligence and occupational status is only .08. The comprehensive analysis of the role of intelligence in occupational status suggests that its influence is primarily indirect by way of the influence of intelligence test score on educational attainment. The model also suggests that the total relation (both direct and indirect) between intelligence and occupational status is of greater strength than the influence of either of the parental variables—father's occupation or father's education. These data suggest that intelligence is not merely a surrogate of parental status but represents a variable that influences ultimate occupational status in American society.

Hauser and Dickinson (1974) have also reported path analysis models of the relation, based on data obtained by Jencks (1972), between intelligence test scores and income. Their analysis suggests that individuals with intelligence test scores in the top fifth of the population of the United States at age 11 will earn about 25% more than the average income and individuals in the bottom fifth of intelligence test scores will earn 19% to 23% less than the average, excluding the influence of family background. Childhood intelligence test score, independent of parental social status, relates to both occupational status and income.

Models such as those presented by Duncan, Featherman, and Duncan should be understood as limited to a particular sample and historical period. Their model is based on data representing white males of nonfarm origin in 1962, who were between 25 and 64 at the time of test. Thus the model does not reflect changed relationships among these variables that may have occurred in the 1960s or 1970s. Also the model may not be valid for other social groups. Duncan (1968) has reported an analysis of occupational status among blacks in 1962. He reports that the relationship between family background and occupational status for blacks is different from that for whites. In particular, blacks of high social status were less likely to be able to pass on their social status to their sons. Jencks (1972) has reanalyzed Duncan's data. Jencks was able to develop a prediction formula for the black sample that related such independent variables as family status and intelligence test scores to social status and income. The typical black person in 1962 came from a family of lower social status than the typical white and had an intelligence test score approximately one standard deviation lower than the typical white person (we shall discuss data on racial differences in intelligence in Chapter 6). Jencks inserted the white mean value into the prediction equation derived for blacks in order to derive a prediction for a mythical black person with the same "advantages" as a white person with respect to these variables. The statistical exercise implies that a black person with the same social background as the average white person in 1962 and the same intelligence test score would have obtained approximately the same amount of education as the typical white person. However, a substantial difference still remained in income. Intelligence differences considered of themselves do not explain or account for status differences between blacks and whites. In 1962, whites with IQ scores of 85 (the presumed black mean) are likely to end up in occupations 12 points below the white mean. Blacks with similar scores end up 24 points below the white mean. At most, cognitive differences account for half the occupational gap. It is not clear if the situation is substantially different today. In any case, these data suggest that the underlying model relating these independent variables to social status and income is different in blacks and whites and that some portion of the gap in social status is not accounted for by mean differences in parental status and differences in scores on intelligence tests.

Intelligence and Success at an Occupation

The relationship between intelligence test score and occupational status has been the subject of some controversy. Herrnstein (1973) has suggested that the relationship between intelligence test scores and occupational status are rational and necessary in any advanced technological society. He asserts:

> The ties among I.Q., occupation, and social standing make practical sense. The intellectual demands of engineering, for example, exceed those of ditch digging. Hence, engineers are brighter, on the average. If virtually anyone is smart enough to be a ditch digger, and only half the people are smart enough to be engineers, then society is, in effect, husbanding its intellectual resources by holding engineers in greater esteem, and on the average, paying them more. The subjective scale of occupational standing that virtually everyone carries around in his head expresses a social consensus both powerful and stable, particularly in its impact on the occupational choices of individuals. It may well be that more people are moved more by that scale than by income, which is merely a correlate of it (and a rather imperfect one at that). More and more these days, young people at the top of the I. Q. scale seem to be choosing the honored occupation, rather than the remunerative one, to the extent that those two aspects can be disentangled. If appearances do not deceive, the correlation between I. Q. and social esteem may be growing even larger than it already is.
>
> The critics of testing say that the correlations between I. Q. and social class show that the I. Q. test is contaminated by the arbitrary values of our culture, giving unfair advantage to those who hold them. But it is no mere coincidence that those values often put the bright people in the prestigious jobs. By directing its approval, admiration, and money towards certain occupations, society promotes their desirability, and hence, competition for them. To the extent that high intelligence

confers a competitive advantage, society thereby expresses its recognition, however imprecise, of the importance and scarcity of intellectual ability [p. 124].[2]

McClelland (1973) has taken a contrary view. He asserts:

Belonging to the power elite (high socioeconomic status) not only helps a young man go to college and get jobs through contacts his family has, it also gives him easy access as a child to the credentials that permit him to get into certain occupations. Nowadays, those credentials include the words and word-game skills used in Scholastic Aptitude Tests. In the Middle Ages they required knowledge of Latin for the learned professions of law, medicine, and theology. Only those young men who could read and write Latin could get into those occupations, and if tests had been given in Latin, I am sure they would have shown that professionals scored higher in Latin than men in general, that sons who grew up in families where Latin was used would have an advantage in those tests compared to those in poor families where Latin was unknown, and that these men were more likely to get into the professions. But would we conclude we are dealing with a general ability factor? Many a ghetto resident must or should feel that he is in a similar position with regard to the kind of English he must learn in order to do well on tests, in school, and in occupations today in America. I was recently in Jamaica where all around me poor people were speaking an English that was almost entirely incomprehensible to me. If I insisted, they would speak patiently in a way that I could understand, but I felt like a slow-witted child. I have wondered how well I would do in Jamaican society if this kind of English were standard among the rich and powerful (which, by the way, it is not), and therefore required by them for admission into their better schools and occupations (as determined by a test administered perhaps by the Jamaican Testing Service). I would feel oppressed, not less intelligent, as the test would doubtless decide I was because I was so slow of comprehension and so ignorant of ordinary vocabulary.

When Cronbach (1970a) concluded that such a test "is giving realistic information on the presence of a handicap," he is, of course, correct. But psychologists should recognize that it is those in power in a society who often decide what is a handicap. We should be a lot more cautious about accepting as ultimate criteria of ability the standards imposed by whatever group happens to be in power.

Does this mean that intelligence tests are invalid? As so often when you examine a question carefully in psychology, the answer depends on what you mean. Valid for what? Certainly they are valid for predicting who will get ahead in a number of prestige jobs where credentials are important. So is white skin: it too is a valid predictor of job success in prestige jobs. But no one would argue that white skin per se is an ability

---

[2] From *I.Q. in the Meritocracy* by R.J. Herrnstein. Copyright 1971, 1973 by Richard J. Herrnstein, by permission of Little, Brown and Co. in association with the Atlantic Monthly Press.

> factor. Lots of the celebrated correlations between so-called intelligence test scores and success can lay no greater claim to representing an ability factor.
>
> Valid for predicting success in school? Certainly, because school success depends on taking similar types of tests. Yet, neither the tests nor school grades seem to have much power to predict real competence in many life outcomes, aside from the advantages that credentials convey on the individuals concerned [p. 6].[3]

One aspect of McClelland's criticism of tests is clearly overstated. That is, as we have seen, intelligence tests reflect something more than social advantages. Indices of social advantage do not account for a major portion of the variance in intelligence test scores. However, McClelland may be right when he suggests that the correlation between intelligence test score and occupational status may not be a necessary feature of a rational society. We may insist on educational credentials for positions where the credentials are irrelevant. Also, the correlation between intelligence test scores and educational success may not be necessary. Perhaps with different admissions criteria and different educational programs, large numbers of individuals with relatively low intelligence test scores could obtain advanced degrees.

The differing views of McClelland and Herrnstein derive in part from different assumptions about the meaning of the relationship between intelligence test scores and occupational success. McClelland argues that the relation is the result of bias and Herrnstein assumes that individuals who score low in intelligence tests would be incompetent if placed in socially prestigious occupations.

What is the relationship between intelligence test score and success within an occupation? On the surface this question appears capable of a simple answer. All that would be necessary to give the answer would be to obtain measures of intelligence and measures of job competence for a group of individuals engaged in the same occupation and report the correlation between these measures. For a variety of reasons such a study is not likely to be definitive. First, one would have to perform the study for a variety of occupations. Presumably, Herrnstein's position implies that the correlation between intelligence test scores and ability to be successful at an occupation would be low or zero in occupations of low prestige. It is only the prestigious occupations that are assumed to require high intelligence and it is only for these

---

[3] From McClelland, D.C. Testing for competence rather than for intelligence. *American Psychologist, 28.* Copyright 1973 by The American Psychological Association. Reproduced by permission.

occupations that one would expect a substantial relationship. However, individuals with low scores on tests are rarely found in such occupations, and so a test of the relationship for individuals with a representative range of intelligence test scores is not possible. Thus the data do not permit us to determine if individuals with low intelligence test scores could successfully be, for example, physicians or accountants. Also, the fact that individuals with high test scores are more likely to enter occupations of high prestige means that individuals with high scores who are in low prestige occupations are a nonrandom sample of individuals with high test scores and this fact may, in unknown ways, influence obtained correlations. In order to definitely discover the relationship between intelligence test scores and occupational competence we would need a society that randomly allocates individuals to different occupational roles without regard to their test scores.

A second set of difficulties stems from the difficulty of obtaining adequate measures of occupational success. It is very difficult to measure the competence of a policeman or a businessman. Frequently, measures of competence are based on judgments that may themselves be contaminated by the raters' sensitivity to intellectual traits combined with the raters' belief (possibly unfounded) about the relationship between intellectual ability and job performance. For example, a trial lawyer with high verbal ability might score high on intelligence tests and be judged highly competent because the person making the judgment believed that verbal ability is positively related to competence. In point of fact, verbal ability might be inversely related or not related at all to competence as defined objectively in terms of the probability of winning jury trials. Jurors might resent the lawyer's verbal ability and, as a result, be inclined to vote against his position.

It should be apparent from the above discussion that it is difficult to obtain a simple answer to the question of the relation between intelligence test scores and success at an occupation. Many writers dealing with this question (see Matarazzo, 1973) quote Ghiselli's summary of research on this issue and assume that the correlation is .20. However, this simple correlation does not do justice to the complexity of the empirical results.

Studies relating intelligence test scores to occupational success can be divided into four categories. First, there are studies of occupations in which only a truncated range of intelligence test scores is found. That is, intelligence test scores below certain cutoff points are exceedingly rare. However, above the minimum level which relates to entry, variations in intelligence test scores are apparently unrelated to success. Typical of such studies is a series of studies by Roe (1953) of extremely eminent scientists. She found a wide range of intelligence test scores in this group with little apparent relationship between eminence and scores on intelligence tests. Similarly, MacKinnon (1962, 1964) has found a nonsignificant negative correlation between scores on intelligence tests and ratings of the creativity of architects. Morsh and Wilder (1954) have reviewed 55 studies relating measures of intellectual ability to measures of teacher effectiveness in the public schools and have concluded that there is no relationship between these variables.

Matarazzo, Allen, Saslow, and Wiens (1964; see also Matarazzo, 1972) have reported a study of the WAIS scores of 243 applicants for police and fire department jobs who had passed civil service examinations. Their scores ranged between 96 and 130 with one deviant score of 86 and a median IQ of 113. Matarazzo (1972) reports after an extensive research program that he was unable to find a relationship between these scores and success as a patrolman.

These data indicate that, in occupations for which high intelligence scores acts as a necessary condition of entry, there is little relationship between occupational success and intelligence test scores.

Second, there are studies dealing with occupations in which there is a relatively wide range of ability represented and where intelligence test scores do relate to occupational success. There are two studies indicating that occupational success in clerical positions is related to intelligence test scores. Hay (1943) related scores on the Otis test of intelligence to an objective measure of success of machine bookkeepers. The criterion of success was the number of transactions completed. He found a correlation of .59 between these measures for 39 female bookkeepers. And, he reports that these results were consistently replicated.

Pond and Bills (1933) reported a study relating occupational advancement among clerical workers to scores on intelligence tests. They reported a correlation of .22 between Otis scores and level of clerical responsibility assigned to an individual at the start of their study. Two and one-half years later the correlation was .41. Individuals who scored high in intelligence tended to advance in responsibility within the clerical ranks.

Studies indicating moderate to substantial positive relations between occupational success and measured intelligence are not restricted to clerical tasks. Scott and Clothier (1923) report a correlation of .51 between scores on intelligence tests and the amount of production of operators in clothing shops.

Third, there are studies in which a sample approaching a wide range of ability is measured and does not relate to measures of occupational success. For example, there is evidence that scores on intelligence tests do not relate to success as a salesperson. Kenagy and Yoakum (1925) report a zero correlation between success as a house-to-house salesman and intelligence test scores. Kenagy and Yoakum also report studies indicating negative correlations of −.11 and −.26 between success as a retail salesperson (according to supervision ratings) and scores on intelligence tests. However, a few studies have found low positive correlation between scores on intelligence tests (particularly for more technical types of sales positions) and success. Success as a life insurance salesman, as measured by actual amount of insurance sold, has been found to be correlated with intelligence test scores with *r*s ranging from .24 to .60.

Results indicating a lack of relationship between scores on intelligence tests and measures of occupational success for individuals with a range of intelligence test scores have not been restricted to studies of salesmen. Otis (1920) reported that there was no relationship between Otis scores and the performance of 400 silk mill weavers. Blum and Candee (1941) have reported a lack of relationship between test scores and ratings and production measures of department store packers and wrappers.

Fourth, although there are few if any studies reporting an inverse relationship between scores on tests and occupational success within an occupation, there are studies suggesting that job dissatisfaction and the probability of remaining on a job are inversely related to scores on intelligence tests. Bills (1923) reported on the stability of a large number of clerical workers in jobs of different degrees of difficulty in relation to scores on intelligence tests. In the simplest clerical job category 27% of workers with IQ less than 80 left within a 30-month period. The comparable percentage for individuals with IQ greater than 110 was 100%. In the clerical position of greatest difficulty the turnover for individuals with low test scores (<80) was 66% and the comparable rate was 41% for individuals of high scores. Similarly, Starch (1922) found that waitresses with low intelligence test scores were much more likely to remain in their jobs than waitresses with high scores.

Viteles (1924) in a study of the stability of cashiers and wrappers as a function of intelligence test scores found that individuals with low scores and high scores were least likely to stay on the job.

It is apparent that there is no simple or easy way of summarizing the relationship between intelligence test scores and occupational success. It is apparent that the relationship between test scores and a person's occupation is simpler and more clear-cut than the relationship between test scores and success of individuals engaged in the same occupation. For occupations of high status, the correlation between intelligence test scores and occupational success approaches zero. However, the severe restriction of range of scores present suggests that there may be a substantial relationship between test scores and performance if individuals with low scores would be allowed to enter the occupation. For occupations of low prestige involving routine work with little intellectual demand, there may be little relationship between test score and success at the job, although there may be a substantial negative relationship between job satisfaction and probability of remaining in the job. Finally, for some jobs, typically with middle levels of occupational status for which low test scores do not act as a barrier for entry into the occupation, the relationship between test scores and performance may be substantial—e.g., for clerical positions. For other jobs in which there is a large range of ability scores present, success may be determined by characteristics other than intelligence scores—e.g., salespeople.

It should be noted that any attempt to summarize these studies is difficult. Much of this literature is dated and good studies are rare. Also, the literature does not directly provide guidance on some of the critical questions about social policy that are implicit in the contrasting views of the importance of intelligence tests as screening devices. It is clear that scores on intelligence tests or other ability measures are frequently used as a means to prevent individuals from entering various occupations. Consider the relation between test scores and entry into positions such as police officer and teacher. In his discussion of research on the relationship between WAIS scores and success as a patrolman, Matarazzo emphasizes his inability to find any relationship between test scores and success as a patrolman for individuals whose scores are **above a minimum** which acts as a cutoff for entry. Note that the potential patrolmen who are given these tests must pass civil service examinations and oral interviews. Matarazzo assumes that the initial screening process, which obviously relates to ability to score well on intelligence tests, is justified. Implicit in this position is the belief that scores on intelligence tests below a certain point would be predictive of failure to succeed as a patrolman. The available data do not provide any guidance on this question. There is a tendency in some police departments to increase the educational requirements for job entry to that function. We can reasonably assume that the consequences of such upgrading will be to raise the average intelligence test score of individuals entering that occupation. Also, it will be more difficult for individuals who tend to score low on tests to enter the occupation. Is this rational social policy? In a sense the answer depends on data that do not exist. Would individuals who score low on intelligence tests be able to perform adequately as patrolmen? Since the data indicate that there is no relationship between measured intelligence and measures of occupational success as a patrolman for a relatively wide range of ability scores, it is not irrational to extrapolate these findings and to suggest that it is possible that individuals who score in the 70s and 80s on tests of intelligence would be just as likely to be successful patrolmen as individuals who score at the median of 113. It does appear to us that any exclusionary use of intelligence test scores should be justified by relevant empirical data. Current attempts to increase minority group membership on police forces in big cities are occasionally retarded by the alleged inability to find a sufficient number of minority group members with the right educational credentials and/or with appropriately high paper-and-pencil test scores. However, since the available data do not indicate that such measures have predictive validity for the job, it is hard to see why they should be used as a basis for exclusion. At the very least, a pilot project could be developed that would include individuals with low scores to discover whether such individuals would be able to perform adequately on the job.

Intelligence and Accomplishment

The relation between test scores and occupational success is only one aspect of the controversy relating test scores to measures of achievement outside of success in school. In particular, it has been suggested that intelligence tests are not measures of creativity and of important real life accomplishment outside of school success. Much of this criticism has come from researchers who have accepted Guilford's distinction between convergent and divergent thinking (see Chapter 2) and who have assumed that divergent thinking measures are measures of creativity and that convergent thinking measures which have a single correct answer are not measures of creativity (Getzels and Jackson, 1962; Torrance, 1962). The issue raised by this critique that is relevant to this chapter concerns the relationship between scores on intelligence tests and accomplishments of importance outside the classroom situation. Research that is relevant to this issue deals with the relationship between scores on intelligence tests and interests and success in extracurricular activities. Wallach and Wing (1969) asked a group of college freshmen to indicate whether they had been involved in each of 34 different activities ranging from participation in extracurricular student organizations to such activities as winning prizes for artwork, debating, or musical accomplishment. The activities covered such diverse fields as art, social service, literature, dramatic arts, music, and science. They found little relationship between an individual's self-report of engaging in such activities and the Scholastic Aptitude Test of verbal and mathematical ability. The percentage of individuals scoring in the highest third of the SAT who endorsed each item was compared with the percentage of individuals in the lowest third of SAT scores endorsing each item. They found significant differences in percentage of endorsement between these two groups for 4 of the 34 items. Individuals in the "low intelligence" group were somewhat **more** likely to have been elected president or chairman of a student organization than individuals in the "high intelligence" group (55% versus 44%) and individuals in the low intelligence group were somewhat more likely to have worked on the editorial staff of a paper or annual than individuals in the high intelligence group (48% versus 36%). Individuals in the high intelligence group were slightly more likely to be appointed an officer of a community or religious group and to have attended an NSF summer science program than individuals in the low intelligence group. On the whole, Wallach and Wing's data indicate that there is no relationship between academic ability and accomplishment outside the classroom. Similar results were reported by Holland and Richards (1965). Wallach and Wing's study dealt with college freshmen and Holland and Richard's study dealt with college students and with college-bound high school students. Obviously, the samples are truncated with respect to intellectual ability measures and provide little evidence of the relationship between intelligence test scores and accomplishment in samples not selected for intelligence. Kogan and Pankove (1974) obtained IQ scores and scores for divergent thinking ability in two schools with a predominantly middle class student body (one school had a smaller student body than the other). In addition they obtained information about the set of extracurricular activities engaged in by students at the time of their graduation.

In the smaller school system, they report a correlation of .66 between intellectual measures in Grade 5 and number of accomplishments outside of school in Grade 12. The comparable correlation for the larger school was .16. The correlation between intellectual ability in Grade 10 and number of accomplishments in Grade 12 was .36 in the smaller school and .39 in the larger school. The data taken as a whole indicate that there is considerable relationship between these measures. That is, in a group that represents something closer to a more typical range of intellectual ability scores, such scores are likely to be predictive of nonacademic accomplishments.

The Wallach and Wing and Kogan and Pankove studies deal with student accomplishments. Of somewhat greater interest in discovering whether intelligence tests are related to real life accomplishments is the relationship between scores on tests and accomplishments in adulthood. There are no data known to us that provide an answer to this question. Ideally one would like to have a study that obtains intelligence test scores on a wide range of individuals while they are in school and then obtains information about their accomplishments (i.e., information other than that of their occupational success). There is, however, one classic study that provides some information about this question. In the 1920s Terman (1925) set out to find a group of children with very high IQs who attended the California public schools. He restricted his search to large school districts (predominately urban) and asked teachers to nominate the brightest children in their classes. Then, these nominees were given a group intelligence test and those who scored above 90% (95% in some cases) were retained for further study. Those pupils were then given a brief version of the Stanford-Binet and those children whose IQs were greater than 130 were than given the full Stanford-Binet. All children under 11 with IQs above 140 were retained for inclusion in the full study. For children between 11 and 14, slightly lower IQs were permitted because of the lack of ceiling in the test. In this way they were able to select 643 subjects with a mean IQ of 151. Only 22 of their subjects had IQs less than 140. It should be noted that the use of teacher nominations undoubtedly biases the selection and some pupils who would have scored sufficiently high for inclusion in the study were never given the tests. Terman was aware of this problem. He reports a study in which the group tests were given to all children in a school in order to see how many children were being missed by the nomination procedure. His data indicate that approximately 10% to 25% of the children who belonged in his group were missed. These children would of course be somewhat different from those included since they were children of high intelligence whose ability was not as apparent to their teachers.

Terman also included a number of siblings of those selected and a group of children who were selected on the basis of group tests only or who were tested by other examiners. The final total group studied included 1528 individuals. Surveys were taken between 1950 and 1955 for this group. Terman and Oden (1959) summarize the results of their surveys and interviews. Although the majority of the women in their sample were housewives and did not choose to pursue a career, they report the following accomplishments for the 700 women they studied: 7 were listed in *American Men of Science*, 2 in the *Directory of American Scholars*, and 2 in *Who's Who in America*. The group had published 5 novels, 5 volumes of poetry, 32 technical or scholarly books, 50 short stories, 4 plays, more than 150 essays, and more than 200 scientific papers. The men's accomplishments are summarized as follows:

> A number of men have made substantial contributions to the physical, biological, and social sciences. These include members of university faculties as well as scientists in various fields who are engaged in research either in industry or in privately endowed or government-sponsored research laboratories. Listings in *American Men of Science* include 70 gifted men, of whom 39 are in the physical sciences, 22 in the biological sciences, and 9 in the social sciences. These listings are several times as numerous as would be found for unselected college graduates. An even greater distinction has been won by the three men who have been elected to the National Academy of Sciences, one of the highest honors accorded American scientists. Not all the notable achievements have been in the sciences; many examples of distinguished accomplishment are found in nearly all fields of endeavor.
>
> Some idea of the distinction and versatility of the group may be found in biographical listings. In addition to the 70 men listed in *American Men of Science*, 10 others appear in the *Directory of American Scholars*, a companion volume of biographies of persons with notable accomplishment in the humanities. In both of these volumes, listings depend on the amount of attention the individual's work has attracted from others in his field. Listings in *Who's Who in America*, on the other hand, are of persons who, by reasons of outstanding achievement, are subjects of extensive and general interest. The 31 men (about 4%) who appear in *Who's Who* provide striking evidence of the range of talent to be found in this group. Of these, 13 are members of college faculties representing the sciences, arts and humanities; 8 are top-ranking executives in business or industry; and 3 are diplomats. The others in *Who's Who* include a physicist who heads one of the foremost laboratories for research in nuclear energy; an engineer who is a director of research in an aeronautical laboratory; a landscape architect; and a writer and editor. Still others are a farmer who is also a government official serving in the Department of Agriculture; a brigadier general in the United States Army; and a vice-president and director of one of the largest philanthropic foundations.
>
> Several of the college faculty members listed in *Who's Who* hold important administrative positions. These include an internationally

known scientist who is provost of a leading university, and a distinguished scholar in the field of literature who is a vice-chancellor at one of the country's largest universities. Another, holding a doctorate in theology, is president of a small denominational college. Others among the college faculty include one of the world's foremost oceanographers and head of a well-known institute of oceanography; a dean of a leading medical scool; and a physiologist who is director of an internationally known laboratory and is himself famous both in this country and abroad for his studies in nutrition and related fields.

The background of the eight businessmen listed in *Who's Who* is interesting. Only three prepared for a career in business. These include the president of a food distributing firm of national scope; the controller of one of the leading steel companies in the country; and a vice-president of one of the largest oil companies in the United States. Of the other five business executives, two were trained in the sciences (both hold Ph.D.'s) and one in engineering; the remaining two were both lawyers who specialized in corporation law and are now high-ranking executives. The three men in the diplomatice service are career diplomats in foreign service.

Additional evidence of the productivity and versatility of the men is found in their publications and patents. Nearly 2000 scientific and technical papers and articles and some 60 books and monographs in the sciences, literature, arts, and humanities have been published. Patents granted amount to at least 230. Other writings include 33 novels, about 375 short stories, novelettes, and plays; 60 or more essays, critiques, and sketches; and 265 miscellaneous articles on a variety of subjects [pp. 146–147].

Terman and Oden's study leaves little doubt that scores on IQ tests do relate to accomplishments outside of academic success. It is doubtful that the attempt to select children scoring in the top 1% of any other single characteristic would be as predictive of future accomplishment.

5

Determinants of Scores on Tests of Intelligence

Genetics and Intelligence

Psychologists for the past 50 years have attempted to use the various methods developed by population geneticists to estimate the relative contributions of nature and nurture to scores on intelligence tests. The estimates that are derived vary depending on the mathematical models used, estimates of parameters in the models which can only be crudely estimated, and the particular data used as a basis for the estimate.

Much of this effort has been directed toward the development of an estimate of heritability usually symbolized as h^2. What h^2 represents is the percentage of variance in some phenotypic characteristic in a population that is attributable to variations in the genotypes possessed by individuals belonging to that population. If h^2 for intelligence test scores is equal to .80 (as is frequently alleged) this implies that 80% of the variance in those test scores for some defined population is attributable to differences in genotypes. It should be noted that h^2 is not a characteristic of a trait—that is, whatever the estimate of h^2 is for intelligence test scores, it is not an estimate that is a property of the phenotypic trait intelligence test scores. Rather, it is a property of the population from which the test scores were obtained. It is extremely important to insist on this apparently esoteric distinction.

Values of h^2 derived for a particular population may not be correct estimates for other populations with different genotypes experiencing different environments. If a value of h^2 was a value for a trait (rather than a population), then one would expect that the value would be invariant for different groups of individuals and that it would be fixed irrespective of variations that might occur in the environment to which groups of individuals might be exposed. However, if h^2 is correctly thought of as a population parameter, then its value may change when individuals are exposed to new environments.

If, for example, it would be possible to reduce the variance of environmental influences that affect intelligence test scores, perhaps through the provisions of optimal educational facilities for all people, then h^2 would go up because a larger portion of variance in the scores would be determined by genotype. Alternatively, if psychologists discovered methods for creating alterations in the environment that would lead to increases in the test scores of individuals who, under present environmental procedures, would tend to score low on intelligence tests and, in addition, if these environmental variations had minimal influence on the scores of individuals who under current environmental conditions tend to score high, then the value for h^2 would decrease because genotype would become a less critical determinant of test scores.

It is an egregious error to derive conclusions about the validity of a program of environmental intervention from estimates of h^2. Even if h^2 were 1.00, it is still theoretically possible that environmental interventions would have major influences on intelligence test scores.

In addition to attempts to derive estimates of h^2, the methods of quantitative population genetics have also been used to obtain estimates of the mode of genetic transmission including estimates of dominance and epistasis. Except for the occurrence of certain rare "genetic errors"—e.g., phenylketonuria, an inherited metabolic disease which usually leads to low intelligence test scores unless special diets and remediation are introduced early in development—scores on intelligence tests have generally been assumed to result from the influence of many genes. That is, a polygenic model has been assumed. Dominance and epistasis refer to the ways in which different genes combine. If dominance exists, the influence of two alleles (genes sharing location on a chromosome, one of which is received from the mother, the other from the father) will not be equal. If dominance does not exist then the combined influence of two alleles will be their average effect. If complete dominance exists then the combined effects of two alleles in the heterozygous case (symbolized by Aa to indicate the presence of one dominant and one recessive allele) will be determined by the dominant allele. If partial dominance exists then the end result will be more strongly influenced by the dominant gene in the heterozygous case.

Epistasis refers to the nonadditive combination of different alleles. If for example, A and B represent genes at different locations that incline an individual to high IQ, if epistasis is present their combined influence might be more (or less) than the sum of their individual influences.

The presence of dominance and epistasis tends to decrease the genetic relationship between a parent and his biological child. By contrast, assortative mating, the term used to describe the tendency of individuals of a particular genotype to mate with individuals having similar genotypes, tends to increase the genetic similarity of a parent and his child because the set of genes received from the second parent is, in some measure, similar or related to the set of genes received from the first parent when assortative mating is present. It is known that the correlation between husband and wife in **phenotypic** intelligence test scores is quite high—approximately .50. Of course the correlation, if any, in their genotypic characteristics that influence intelligence test score can only be estimated. Paradoxically, the estimate depends on derived estimates of h^2 which in turn frequently depend on estimates of the value for assortative mating.

The methods of quantitative population genetics can also be used to derive estimates of the influence of the environment on intelligence test scores. The environmental influences encompassed in such estimates are, of necessity, quite diverse. They include the potential influence of the biological environment including such potential influences as health and nutrition, and the influences of test unreliability including random variations from time to time in performance on a test by the same individual. In addition, environmental variations have been partitioned into two components—those attributable to variations between families and those attributable to variations within families. Just as h^2 represents an estimate of the percentage of variance that genotypes contribute to test scores, e^2 represents the percentage of influence of the variations in a phenotypic characteristic in a particular population attributable to variation in the environment.

In addition to the contributions of h^2 and e^2 to test scores, the methods of quantitative population genetics have been used to derive estimates of the covariance of genotype and environment. It is generally assumed that parents with genotypes that tend to lead to high intelligence test scores are likely to provide their children with environments that are also favorable for the development of intelligence test scores. Similarly, parents possessing genotypes that are assumed to lead to low intelligence test scores are generally assumed to provide for their children environments that are not favorable to the development of high scores on tests. This reasoning implies that there will be a positive correlation between genotype and environment and that the covariance contribution to phenotypic scores will not be zero.

Covariance between genotype and environment may occur for a second reason. Individuals with a genotype that predisposes toward the development of high intelligence test score may choose or select an environment favorable for the development of high scores. For example, a child who is successful in school by virtue of having a genotype that tends to the development of high intelligence test score is likely to enter an academic track. If entering an academic track serves to increase his test performance then he will receive high test scores by virtue of the self-selection of features of the environment that are favorable toward the development of high test scores. This type of covariance may be considered an influence of the genotype on the phenotype—albeit a somewhat indirect influence. However, the correlation not due to self-selection attributed to the assumed tendency of genotype and environment to be correlated is not legitimately assigned to a genotypic influence.

A linear additive model of the determinants of a phenotype may be algebraically expressed as follows:

Phenotypic Score = Genotype + Environment
+ Covariance of Genotype and Environment

The model explicitly assumes that the factors combine additively. It is possible to design a model that assumes nonadditive combinations among the components. However, the additive model is preferred on the grounds of simplicity. And, at the present time there is little evidence that would suggest that a nonadditive model is required or superior. Also, nonadditive sources of variance may be included in the additive model by the addition of a term referring to interaction variance. Interaction would result if certain environments when combined with certain genotypes (and not others) have an unusually favorable or unfavorable influence on the phenotype—an influence different from that of the addition of their general effects. There is little evidence, at present, that interaction variance is a major source of variance in intelligence test scores.

There are at least three different positions on the influence of heredity on intelligence. The "classical" position which has been reaffirmed in recent years by Jensen (1973) and Herrnstein (1973) suggests that approximately 80% of the variance in intelligence test scores is attributable to genetic differences between individuals. Jencks (1972, see especially Appendix A) analyzed data on the heritability of intelligence test scores and has concluded that the best estimate is that genotypic differences explain approximately 45% of the variance in scores. He goes on to say, "in gamblers' terms we think the chances are about two out of three that the heritability of IQ scores, as we have defined the term, is between 0.35 and 0.55, and that we think the chances are about 19 out of 20 that heritability is between 0.25 and 0.65 [p. 315]."

Finally, Kamin (1974) has reviewed the literature on the inheritance of intelligence test scores and concluded, "There exists no data which should lead a prudent man to accept the hypothesis that I.Q. test scores are in any degree heritable." Note that Kamin's conclusion is not that IQ scores are not heritable but rather that the data, which have been interpreted by many psychologists as indicating that the scores are partly determined by genotype, do not support such a conclusion.

In summary, there exist three apparently discordant positions on the influence of genotypes on IQ. We shall present the argument and reasoning put forward by the advocates of each position and then attempt to come to some overall evaluation of the evidence.

The different conclusions reached about the influence of heredity on intelligence test scores are partly due to the use of different methodologies for the analysis of data. More importantly, however, the differences in conclusion are related to differences in the data used to reach conclusions. The "classical" view relies heavily on the data presented by Cyril Burt collected in England. Jencks indicates that Burt's data, as well as some other data collected in England, tend to indicate larger influences for heredity than do the American data. Jencks's analysis is based solely on data collected in America. Accordingly, the disagreement on the magnitude of the influence of heredity on intelligence test scores between the views of Jensen and Herrnstein and those of Jencks is substantially influenced by the use of different data bases for the purpose of analysis. There are also differences in methods of analysis.

The differences between those psychologists who are convinced that intelligence test scores are influenced by heredity on the one hand, and Kamin on the other hand are almost totally related to differences in beliefs about the value of the data used in the analysis. Kamin would agree that the data usually cited in connection with studies of the inheritance of intelligence, **if accepted at face value,** provide clear-cut evidence for the belief that intelligence test scores are largely determined by genetic influences. Indeed, if one accepts Burt's data at face value, environmental influences of any sort are barely discernible. Kamin's analyses deal substantially with the probity and value of the data in this area. He argues that many of the studies contain flaws and methological errors and that when the studies are examined critically, without prior commitment to a hereditarian viewpoint, the conclusions reported in the studies and the usual summaries of the studies are without merit.

In summary, the radically different conclusions about the influence of heredity on intelligence test scores that exist in the literature are principally attributable to differences in the data used to arrive at conclusions and differences in beliefs about the value of the existing data.

It is convenient to consider separately three bodies of data relating to the influence of heredity on intelligence test scores—studies dealing with the relationships between adopted and natural children's test scores and the scores of their parents, studies dealing with relationships in test scores between children who differ in their genetic similarity, and studies of children of identical or similar genotypes who are reared apart.

Studies Dealing with the Relationship between Children and Parents

There are three kinds of data dealing with the relationship between parents and children that are relevant to an analysis of the influence of genotype on intelligence test scores. These include an analysis of differences between the correlation of intelligence test scores between biological parents and natural children brought up by these parents, adopted parents and adopted children brought up from an early age by their adopted parents, and natural parents with their natural children who were adopted and brought up by other parents. If intelligence test scores were completely determined by genetic characteristics we would expect the correlation between the intelligence test score of natural parents and their children adopted from birth to be the same value as the correlation between the scores of natural parents with their children reared in the home of the natural parent. That is, the variations introduced by raising children in a home different from that of their biological parents ought not to influence their scores. The correlation between the intelligence test scores of adopted parents and the intelligence test scores of their adopted children should be zero if scores are determined solely by genotype and if the genotype of the adopted parents for intelligence are unrelated to the genotypes of the natural parent—i.e., there has been no selective placement for intelligence based on genotype.

Jencks reviews studies dealing with the correlation between test scores of children and that of their natural parents. Table 5.1 presents the results of these analyses. Table 5.1 indicates that the studies use somewhat different measures. In particular the deviant value of .35 comes from a study in which Stanford Achievement Tests were included—a measure not usually conceived of as an intelligence test. Note further that Jencks computes mean values that are weighted by sample size. This is, as we shall see, a source of difference in his estimates of heritability.

TABLE 5.1
Observed Correlations between Parents' IQ Scores and Children's Stanford-Binet Scores When Children Are Raised by Natural Parents[a,b]

Source	Parental test	Number of parents	Correlation
1. Burks, *Nature and Nurture*	Stanford-Binet	200	.46
2. Leahy, *Intelligence*	Otis IQ	366	.51
3. Conrad & Jones, *Second Study*	Army Alpha	441	.49
4. Outhit, *Resemblance of Parents*	Army Alpha	102	.58
5. Willoughby, *Family Similarities*	Stanford Achievement, Army Alpha, and NIT	141	.35
6. Higgins, Reed & Reed, *Intelligence and Family Size*	Group tests in school	2032	.44
Weighted mean of samples 1–5		1250	.48

[a]Correlations computed separately in original studies for fathers and mothers. Correlations shown here are means of virtually identical values; N is total for both fathers and mothers.

[b]Based on Jencks (1972).

Jensen and others have relied on median values obtained from separate studies. Actually a valid argument can be made for the use of either procedure. Weighted means are appropriate because they assume that a study with a larger number of subjects contributes more importantly to the estimate of a combined correlation value than a study with a smaller number of subjects. On the other hand, each of the studies whose results are combined uses somewhat different methodologies and involves different samples. From this point of view some studies may be "better" or more representative than others for reasons quite independent of sample size. If this were true, then a procedure that in effect considered each study equally would be justified. The difference between the use of medians or means is somewhat less critical. Means weight extreme values derived from a study more than medians. The choice between them hinges on the interpretation of extreme scores as either being due to sampling variations or as reflecting errors due to aberrant methodological procedures.

In the case of the correlation between the test scores of the natural parents and their children reared by them, Jencks's values do not differ from those derived from the median of international studies reported by Erlenmeyer-Kimling and Jarvik (1963). Jencks assumes a correlation of .48 whereas Erlenmeyer-Kimling and Jarvik assume a correlation of .50. Jencks proceeds to correct the value of .48 in two ways. First, he notes that each study deals with a sample from a limited geographic area and thus represents a curtailment in the influence of the environment sampled. He assumes a national sample would yield a correlation value of .505. Second, he corrects the correlations for attenuation or unreliability and attains a correlation of .55. The correction for attenuation, which is used everywhere in Jencks's work, tends to inflate the estimate of heritability because unreliability in test score must be an environmental influence.

The correlation between Stanford-Binet intelligence test scores of adopting fathers and their adopted children ranges in three American studies summarized by Jencks between .07 and .37 with a weighted mean value of .21. The corresponding correlation from the same studies of adopting mothers with their adopted children ranges between .19 and .24. Jencks uses a value of .225 to represent the estimated correlation between adopting parents' scores and the scores of their adopted children. Correcting this value for unreliability and restriction of sample yields a corrected value of .28. The difference between the value of .55 for the correlation between natural parents' test scores and the scores of their children and the value of .28 suggests that there is some influence of genotype on test score. (Alternatively, these data, taken by themselves, are compatible with the somewhat improbable view that all of the parental influence on test scores occurs during the prenatal period, the perinatal period, and in the first few months after birth where the child to be adopted is still under the influence of the biological parent.)

In order to use these correlations to estimate heritability, two additional estimates are required. First, some indication of the effects of selective placement in the adoption studies is necessary. This is probably not of great importance because the correlation between the mean IQ of biological parents and the mean IQ of the adopting parents is only .20. Hence the genotypic relations between them is not likely to be large. A more important issue involved in solving the path equation deals with the genetic relationship between parents and their children. It is usually assumed that the genetic relationship between a child and one of his biological parents is .50 because a child receives half of his genes from each parent. However, this is an oversimplification. The genetic correlation between a child and his parent depends on the way in which the genes combine. If dominance exists, the influence of this pair can be dependent on one of the alleles (the dominant one in the Aa case). If partial dominance exists then this pair will be closer to the partially dominant allele. In addition to the effects of dominance, epistasis may exist—e.g., two favorable alleles at different loci might lead to higher IQ than would be expected by adding their combined effects. Finally, the genetic correlation between parent and child is influenced by assortative mating.

Jencks estimates, on the basis of a weighted mean of nine U.S. studies, that the correlation between intelligence test scores of husband and wife is .50. Assortative mating suggests that there may be some correlation between the genotype for intelligence between the parents and hence the genetic influence of one parent may not be unrelated to the genetic influence of a second parent.

Jencks uses these data to estimate h^2. His estimates vary with the value assumed for the genetic correlation between parent and child (y) which is based on estimates of the effects of dominance and epistasis. If one assumes that $y = .50$ then Jencks's path analyses imply that $h^2 = .29$, e^2 (the proportion of variance attributable to the environment) = .52, and the covariance due to the correlation of heredity and environment is equal to .19. If y is as low as .26, Jencks's analyses imply that h^2 is equal to .76, $e^2 = .06$, and the covariance is equal to .18.

Jencks's analysis of correlation between parents and children suggests first that the value of h^2 is a function of the estimated genetic relationship between parent and child. Second, there is under any assumption of that value a substantial influence of covariance between genetic and environmental factors. The covariance derives from the assumed relationship between a child's genotype and a child's environment. That is, children with presumably favorable genotypes for the development of intelligence are likely to be born into families that provide favorable environment for the child.

Jencks's analysis of data on the relationship between children's intelligence test scores and the intelligence test scores of their natural or adopted parents overlooks an important issue central to Kamin's discussion of these data. Adopted parents represent a biased sample. In the studies reported, they tend to be older than natural parents having children of the same age. Their adopted children tend more often to be only children. Their adopted children are more likely to be female and adopted parents tend to have been selected rigorously by social agencies and tend to be individuals with good "reputations." Thus, to a certain extent adopted families are homogeneous. The reduction in variance on variables that have a potential influence on intelligence test score would, on purely statistical grounds, tend to lower the correlation between adopted parents' test scores and those of their adopted children on the reasonable supposition that the between-family variance on environmental characteristics with some potential influence on intelligence test scores is less for adopted families than for natural families. And, this possible restriction in range of talent would account for a decrease in the correlation between parent and child in adopted families.

Data exist that permit one to circumvent this problem. The correlation between the intelligence test scores of parents who have adopted a child and the intelligence test scores of the biological children of such parents should provide a relevant basis for comparison with the correlation between intelligence test scores of adopted children and their adoptive parents. Kamin's summary of these data is presented in Table 5.2. Table 5.2 indicates that the relationship between the test scores of adoptive parents and their biological children is lower than the usual correlation reported between natural parents and their children reared by them. The correlation of .57 for "control" children derives from data in the studies of Burks (1928) and Leahy (1935) contrasting the relationship between adopted children and their adoptive parents with those of a group of natural children and their natural parents where the control group for

TABLE 5.2

IQ Correlations from Studies of Adopted Children[a]

Study	Adopted child × Adoptive midparent	Own child × Adoptive midparent	Control child × True midparent
Freeman *et al.* (1928)	.39 (N = 169)	.35 (N = 28)	—
Burks (1928)	.20 (N = 174)	—	.52 (N = 100)
Leahy (1935)	.18 (N = 177)	.36 (N = 20)	.60 (N = 173)
Pooling all studies	.26 (N = 520)	.35 (N = 48)	.57 (N = 273)

[a] Based on Kamin (1974).

natural parents was matched principally on indices of occupational characteristics of the fathers. The data in Table 5.2 indicate that the relationship between the test scores of adoptive parents and their adopted children does not differ substantially, if at all, from the relationship between adoptive parents and their natural children (r = .26 versus .35). However, this latter correlation is based on a very small group of children (N = 48). Although the N is small, the group of 48 provides a more relevant comparison group and the correlation of .35 is a more appropriate correlation than an estimated correlation for the relationship between parents in general and their biological children. Kamin's analysis therefore has the virtue of indicating difficulties in the usual analysis of these data and pointing to an important additional datum that has heretofore not been accorded sufficient weight in discussions of these data.

As a result, it is not unreasonable to conclude that existing data involving comparisons between the relationship of adopted parents and their adopted children and biological parents with their biological children in intelligence test scores do not permit any firm conclusion about the influence of genotype on intelligence test score.

It should be noted that Jencks's path analysis of relationship between parents and children does not include data on the correlation between natural parents and their biological children who were adopted and reared apart. Jencks's analysis unfortunately is based on only one study dealing with this relationship. Skodak and Skeels (1949) reported a study of adopted children in which they found that the correlation between the intelligence test scores of natural mothers with their biological children adopted at at early age increased as the child became older. The correlation rose to a value of .41 which, when corrected for attenuation, is .446, and is in excess of that expected by any of the values derived from the analysis of correlations between biological parents and children reared in the home and adopting parents and children. However, Jencks cites only one study of this relationship (i.e., the relationship between the intelligence test scores of biological parents with the scores of their adopted children). Jencks assumes, in light of the small number of cases, that the corrected value of .446 is artificially inflated because of sampling errors. If the value is replicated, this would suggest that Jencks's estimates of heritability derived from these data are too low. Loehlin, Lindzey, and Spuhler (1975, Appendix I) have criticized some aspects of the mathematical solution of the path equations used by Jencks to derive heritability estimates from these data. They point out that certain minor but plausible changes in mathematical procedure would increase h^2 values for the data used by Jencks and would, in addition, lead to expected values of the correlation between biological mothers and their children adopted from birth that are in line with those reported by Skodak and Skeels.

Skodak and Skeels also reported that the correlation between the biological mothers' educational level and the intelligence test scores of their adopted children was .32. The comparable correlation at an average age of 13 between the adopting mothers' educational level and the test score of their adopted children was .02. This finding suggests that the outcome of the child's intelligence test score is determined principally by the characteristics of his biological mother rather than by the characteristics of the adoptive mother (Honzik, 1957). Their finding is frequently cited by individuals who wish to argue for the dominance of nature over nurture in the determination of scores on intelligence tests. Kamin has a number of criticisms of this interpretation of the Skodak and Skeels study. First, he indicates that these correlations are based on 100 cases who were the survivors of a group of 180 children originally selected for study. However, data for the 139 children available for testing at age 7 indicated that the biological mother's educational level correlated with her biological child's intelligence test score .24. The comparable value for the correlation between the educational level of the foster mother with her foster child was .20. These values do not provide anywhere near the dramatic evidence for the dominance of genetic influence over that of the influence of the environment. A person favoring a genetic interpretation might argue that the increases in the value of the correlation between the biological mother and the child and the decreases in the value of the correlation between the foster mother and her child were due to the "maturing" influence of the genotype on the phenotype. This argument is somewhat strained because intelligence test scores at age 7 are reasonably predictive of subsequent intelligence test scores and should, as a result, reflect the influence of the genotype. Kamin prefers to explain the increased correlation between the characteristics of the biological mother and the child and the subsequent decrease in the correlation between the foster mother and the child by reference to changes in the composition of the sample of children and mothers available for study. He points out that 51% of the 100 children available for study at age 13 had foster mothers who had attended college, but 29% of the 38% of the mothers who had dropped out of the study had attended college. This suggests that the sample became slightly more educated. And, Kamin argues that the increasing homogenization of the sample in the educational level of the foster parents who remained in the study decreased the correlation between their educational level and that of their foster children.

Kamin's argument on this point seems strained. Although he does not note this, the sample of 138 tested at age 7 was one in which 46% of the sample of foster mothers had attended college. This does not appear to be a dramatic difference. Also, a sample in which 52% of the mothers attend college (but do not necessarily graduate from college) is not one that has a sharply restricted range of educational levels among the foster mothers. That is, educational levels of this degree of variance would probably correlate positively with children's intelligence in most samples.

Kamin's interpretation of the positive correlation of biological mothers' intelligence with the test scores of their children who were adopted appeals to selective placement. The difficulty with this argument is that one would have to have substantial positive correlation between the characteristics of the biological mother and the foster mother to obtain correlations of the order of magnitude reported by Skodak and Skeels between characteristics of biological mothers and their adopted children. While the available data do support the idea of selective placement, the correlation between characteristics of the biological mothers and the foster mothers who adopted their children are of relatively small magnitude.

Kamin has noted one rather bizarre feature of the Skodak and Skeels data. The correlation between the educational level of biological mothers and their adopted daughters was .44. The same correlation for boys was −.01. This difference, which is statistically significant, does not readily fit a genetic interpretation.

Additional data relating characteristics of biological parents to their adopted children would certainly be desirable. Munsinger (1975) has surveyed all the available literature relating the IQ of the adopted child to characteristics of the biological parent and the adopted parent. His review includes the results of some recent work that is, in some respects, superior to the earlier literature on this topic. Two recent studies are of particular significance. Horn, Loehlin, and Willerman (as cited in Munsinger, 1975a) have reported a correlation of .32 between the IQ score of 191 biological mothers and the IQs of their biological children released for adoption after delivery. This correlation is not attributable to the effects of selective placement because the correlation between the IQ of the biological mother and that of the adoptive mother was only .18, and the correlation of the adoptive mother's IQ with that of the adopted child's IQ was .15.

Munsinger (1975b) has reported correlations between IQ scores of adopted children at an average age of 8.5 years and the educational and social status of their biological parents for a group of children adopted shortly after birth. For a sample of 41 children, he reports a correlation of .70 between the biological midparent social background (i.e., the average of the midparent social score) and the IQ of their biological child. The corresponding correlation between the adoptive midparent social background and that of the adoptive child was −.14. There was no selective placement in this sample. It should be noted that the correlation between the biological midparent social background and the IQ of their children adopted after birth is higher than expected and is higher than the usually reported correlation between the midparent social background and the IQ of children reared by their biological parents. The high correlation in this study is most parsimoniously explained as being due to sampling error in a relatively small sample.

These studies support the findings of the Skodak and Skeels study and indicate that the IQ of children adopted from birth is related to the intellectual and social background of their biological parents, and, further, that this relationship is not attributable to the effects of selective placement. This latter conclusion is buttressed by two additional findings—the correlation between the biological parents' characteristics and those of the adoptive parents is invariably low and the correlation between the adoptive parents' characteristics and the intellectual level of their adoptive children is also relatively low. Munsinger's careful summary of all of the available data leads him to the conclusion that the best estimate of the obtained correlation in several studies between the social and intellectual characteristics of the biological midparent and the intelligence test score of their biological children released for adoption is .48. This value should be compared with his average of obtained correlations of .58 for the biological midparent correlation and the intelligence test score of their children reared by them. Munsinger's analysis leads to two conclusions. First, Jencks's path model cannot accommodate these data. That is, the data on the relationship between the intelligence test score of adopted children and the intelligence and social backgrounds of their biological parents suggest that h^2 for intelligence test scores must be higher than Jencks believes. Second, these data cannot be explained without appeal to a substantial influence of genotypes on intelligence test scores. In this respect Kamin's critique of the Skodak and Skeels study cannot be extended to some of the newer research.

Relationships among Children Reared Together Who Differ in Genetic Similarity

Studies comparing children reared together who differ in their degree of genetic similarity may be used to estimate heritability of intelligence test scores. The basic assumption underlying this method of determining the influence of heredity is that children reared together who resemble each other genetically ought—on the assumption that genetic characteristics influence test scores—to obtain similar test scores. In addition, the degree of similarity should be determined by the degree of genetic resemblance. Jensen (1967) has provided a simple formula for estimating heritability from such data:

$$h^2 = \frac{r_{12} - r_{34}}{g_{12} - g_{34}},$$

where r_{12} refers to the correlation on the trait between a pair reared together who are genetically similar, and r_{34} is the correlation between pairs reared together who are relatively less genetically similar; g_{12} is the genetic correlation between the first pair, and g_{34} the genetic correlation between the second pair.

This formula may be applied to studies comparing the relationship between monozygotic twins (identical twins, hereafter MZ) same-sex dyzygotic twins (fraternal twins, hereafter DZ), and siblings and unrelated children reared together. Monozygotic twins are assumed to have a genetic correlation of 1. The genetic correlation of DZ twins depends, in part, on assumptions about influence of assortative mating and dominance and epistasis which tends to affect the value of genetic similarity for siblings in general and DZ twins in particular. A commonly used value for genetic similarity among siblings is .55.

It has long been known that MZ twins resemble each other more than DZ twins in intelligence test scores. The precise degree of resemblance in scores varies from study to study and is related to the test used and to the sample, as well as to chance fluctuations.

There are a number of American studies in the literature providing data on the relationship between MZ and DZ twins on intelligence that use adequate tests of zygosity and individually administered tests of general intelligence. Newman, Freeman, and Holzinger (1937) report correlations of .91 and .64 for 50 pairs of MZ and 50 pairs of DZ twins attending Chicago public schools. Jencks reports data obtained by Schoenfeldt for an "IQ composite" derived from a group test given to a large number of public school students in the U.S. indicating an MZ correlation of .85 for 335 cases and a DZ correlation of .54 for 156 cases. Also, there are a number of English studies that have reported correlation on a variety of group and individual tests for MZ twins (Jencks, 1972), ranging from .76 to .94 for MZ twins, and values ranging from .44 to .66 for DZ twins with weighted means calculated by Jencks of .87 for 211 MZ twins and .51 for 412 DZ twins. In addition there have been some large-scale studies reporting correlation between DZ twins. Mehrota and Maxwell (1950) found correlation for a large sample of opposite-sex twin pairs in Scotland of .63; Husen (1963) reported a correlation of .70 for 416 Swedish males taking a test for military induction. Erlenmeyer-Kimling and Jarvik (1963) report a median of .53 for nine studies reporting the relationship among DZ twins. It is this latter value (or one similar to it) that has been used by psychologists such as Jensen, Herrnstein, and Burt, who have assumed high heritability for intelligence. Nevertheless, there is some reason to believe that large-scale representative studies of DZ twins using individual IQ tests might obtain correlations in excess of .53. As the value for DZ twins used in the formula increases, the estimate of h^2 decreases. However, there is no doubt that MZ twins resemble each other in intelligence more than DZ twins do.

Jencks, who cites data on siblings reared together, computes a weighted mean correlation of .52 for seven U.S. studies encompassing 1951 pairs of siblings. Note that the correlation for siblings is less than the correlation for MZ twins, again indicating, for children reared together, that similarity in intelligence test score is determined in part by genetic similarity.

A fourth source of data on similarity of children reared together comes from studies of biologically unrelated children reared in the same house. There is relatively less information available about such children. Jencks summarizes the available American data and reports a weighted mean correlation value of .32 for 259 unrelated children reared together as reported in four studies (none of which was recent). Jencks corrects the value of .32 for unreliability and restriction in sampling range and arrives at a value of .38. The results of substituting the values for MZ, DZ, siblings, and unrelated children reared together into Jensen's simplified formula yield h^2 values ranging from .40 to .84. It can be seen that there is considerable variability in estimated h^2 values. However, in each case there is evidence that children who are relatively more genetically related and who are reared together are more likely to be similar in intelligence than children who are relatively less genetically related.

KAMIN'S CRITIQUE

Although there is little or no doubt that MZ twins are more alike than DZ twins on intelligence test scores, Kamin points out that this finding is not critical because MZ twins, by virtue of the similarity of their appearance, are more likely to experience a similar environment than DZ twins. He cites data reported by Wilson (1934; see also Smith, 1965), indicating that MZ twins are exposed to more similar environments than DZ twins. For example, MZ twins are more likely to have similar friends (in Smith's study, 58% of MZ, compared to 33% of DZ twins, reported similar friends). Forty percent of Smith's MZ twins reported they studied together compared to only 15% of like-sex DZ twins. Data such as these are used by Kamin to argue that the MZ–DZ difference in similarity of intelligence test score is really not critical. However, it should be noted that data such as Smith's do not provide a critical test of the notion that the differences in degree of similarity between MZ and DZ twins are attributable to differences in the degree of similarity of their environment. What is lacking is evidence that the differences in these characteristics have causal importance. There is little hard evidence, for example, that studying together **produces** or **leads to** similar intelligence test scores. That is, greater similarity of experience for MZs does **not** explain greater similarity of test scores unless it can be shown that the particular respects in which MZs experience greater similarity of environment cause greater similarity of scores. It should also be noted that while it is unlikely that MZs and DZs experience equally similar intrapair environments, there are many respects in which they obviously share environmental similarity. Also, the greater similarity of environment experienced by MZ twins as reported in the Smith study may in part be the result of greater genetic similarity for intelligence. Thus, if children select their friends on the basis of intellectual similarity and if MZ twins are more likely to be genetically similar in intelligence, then they are more likely than DZ twins to choose the same friends, who may be genetically less similar in intelligence. A similar argument may be applied to the environmental variable, studying together. Thus, similarity of environment may be in part the result of genetic similarity and not the cause of similarity in test score. Hence the similarities in test scores and in environments may both be produced by genetic similarity. Thus the data on greater similarity of environment of MZ twins than DZ twins does not vitiate a genetic explanation of greater similarity of test score. On the other hand, the findings do at least suggest that data on differences between MZs and DZs do not unambiguously permit inferences about genetic similarity.

The argument that the greater similarity of MZ twins relative to DZ twins is due to the greater similarity of environment experienced by MZ twins has two testable consequences—one of which is noted by Kamin, the other is not. The environmental interpretation would appear to imply that DZ twins are more similar to each other than are siblings. Furthermore, it would not be unreasonable to suggest that an environmental interpretation requires that DZ twins be more similar to MZ twins in IQ resemblance than they are to siblings. This inference is based on the assumption that there are probably important environmental influences shared by twins of any kind that are not shared by siblings.

Kamin cites a number of studies germane to this issue. He reviews studies in which the same investigator using the same methods reports correlations for both DZ twins and siblings. In two of the three studies reported the siblings resembled each other less than DZ twins.[1]

In addition to studies comparing DZ twins with siblings, studies exist in which a member of a DZ pair is compared to his nontwin sibling. On purely genetic grounds, one would expect that a member of a DZ pair would be as similar to his DZ twin as to his nontwin sibling. Table 5.3 presents a summary of studies cited by Kamin on this issue. Three of the four studies indicate that DZ twins are clearly more similar than DZ–sibling pairs. However, only one of the studies used an individual intelligence test and this study reported the greatest discrepancy between these DZ pairs and DZ–sibling pairs. These studies, combined with studies in which both sibling and DZ twin data were reported, apparently suggest that siblings are significantly less similar in IQ scores than DZ twins. This in turn supports Kamin's environmental interpretation of the MZ–DZ difference.

On the other hand, other data exist that are less supportive of this conclusion. Jencks has summarized data that indicate a weighted mean correlation value of .52 for seven studies in which the Stanford-Binet was used, containing, in total, 1951 sibling pairs. While

---

[1] N.B. The third study that reported the same results for DZ twins and siblings, Burt's study, is one Kamin has criticized on other grounds and is one that he discounts.

TABLE 5.3
Correlations between siblings and DZ Twins[a]

Study	Test	*N*	DZ pair	DZ-sib pair
Snider	Iowa test of Basic Skills Vocabulary	329	.50	.26
Stocks & Karn	Stanford-Binet	104	.65	.12
Partanen, Brun, & Markkanen	Verbal comprehension	90	.51	.41[b]
Huntley	Composite vocabulary	108	.58	.58

[a] Based on Kamin (1974).
[b] This study could not be obtained by us. Kamin gives two values and we infer from the order of presentation that these values were for the verbal comprehension factor.

the comparable value for DZ twins cannot be determined with equal certainty, enough additional data are available to suggest that the comparable value for DZ twins is probably in the 60s. Also, we have reason to believe that the reported correlation underestimates the value for genetic siblings because of the probable inclusion of half siblings in the sibling group and the probable influence of imperfect age standardization. Although there is some uncertainty surrounding all of these values, on the whole the available data suggest that the relationships between siblings on individually administered general tests of intelligence is not substantially different from the relationship between DZ twins. On the other hand, studies comparing DZ nontwin sibling pairs with DZ twin pairs suggest an opposite conclusion.

Another source of data that is relevant to Kamin's argument about the environmental basis for the MZ–DZ difference comes from studies comparing same-sex DZ twins with opposite-sex DZ twins. It is reasonable to assume that the intrapair difference in environment of opposite-sex DZ twins is greater than the intrapair difference in environment of same-sex DZ twins. This expectation follows from the sex segregation typical of American society and the probable difference in socialization practices experienced by boys and girls. Furthermore, it is not unreasonable to assume that the difference in intrapair similarity of environment between opposite-sex DZ pairs and same-sex DZ pairs is likely to be larger than the differences in similarity of environment between same-sex DZ pairs and same-sex MZ pairs. If this line of reasoning is correct, an environmental hypothesis to explain the MZ–DZ twin difference in similarity of intelligence test scores would predict that same-sex DZ twins should be considerably more similar to each other than opposite-sex DZ twins. Many of the available twin studies have reported data for same-sex MZ twins and same-sex DZ twins. And, some studies have pooled same-sex DZ and opposite-sex DZ twins into a single category. The available data provide little or no support for the expectation that same-sex DZ twins are significantly more similar to each other in intelligence test scores than opposite-sex DZ twins. Erlenmeyer-Kimling and Jarvik's summary of studies reports median values for opposite-sex DZs and same-sex DZs of .49 and .56, respectively.

Mehrota and Maxwell (1950) studied virtually the total sample of Scottish 11-year-olds and obtained correlations of .63 for 182 opposite-sex DZ twins. In another study cited by Kamin, Herrman, and Hogben (1933), correlations are reported of .47 and .51 for 96 and 138 same-sex and opposite-sex English DZ twins on the Otis test. The available data accordingly support the view that opposite-sex DZ twins and same-sex DZ twins do not substantially differ in their degree of resemblance in intelligence test scores. This comparison is, in some respects, less confounded than the comparison between DZ twins and siblings. There is no chance that opposite-sex DZ twins are mistakenly classified as full siblings when they are half siblings. Problems in age standardization of test scores are not involved in the comparison. Another source of bias possibly present in twin studies is also controlled in this comparison. Kamin has indicated that MZ twins and DZ twins may be drawn from somewhat different strata of the population. For example, the probability of having a DZ twin tends to increase with age. Accordingly, DZ twins tend to be born to older mothers. This sampling (along with others indicated by Kamin—e.g., overrepresentation of female DZ pairs in the available literature) may influence the comparison of DZ and MZ twins. However, comparisons **among** DZ twin pairs are less likely to be subject to possible sampling bias. There is, however, one major source of bias in the comparison between same- and opposite-sex DZ twin pairs. To the extent to which there are errors in the determination of zygosity, they are restricted to the same-sex DZ pairs. And, since the available data clearly indicate that MZ twins reared together are quite similar in intelligence test scores, any errors of zygosity determination are likely to inflate the similarity of same-sex MZ twins in test scores relative to the similarity of same-sex DZ twins. Since the available data suggest that same-sex DZ twins and opposite-sex DZ twins do not substantially differ (if they differ at all) in resemblance in intelligence test scores, we can infer that differences within family environments of twins are not important contributors to their degree of resemblance in intelligence test score. This tends to argue against Kamin's environmental interpretation of the MZ–DZ difference in similarity of intelligence test score.

Studies of Genetically Related Children Reared Apart

The most important class of studies of this type deals with MZ twins reared apart. If MZ twins are reared in different environments their degree of resemblance provides an estimate of the degree to which identity of genotypes leads to similarity of test scores under different environmental conditions. Confirmed hereditarians have sometimes used the correlation between separated MZ twins as a direct estimate of h^2. This procedure presupposes, among other things, that the environments in which the twin pairs are raised are substantially uncorrelated. Relationships among MZ pairs reared apart are also germane to the environmental interpretation of the difference between MZ and DZ twins reared together. Unfortunately, it is exceedingly difficult to find MZ twins reared apart. The entire published literature on this issue consists of four studies containing in total 122 MZ twin pairs. The obtained correlations reported in the literature range from .69 to .86 with a weighted mean of .81. These data, taken at face value, suggest that MZ twins reared apart are quite similar with respect to intelligence and appear to be perceptibly more similar than DZ twins reared together. These data have been used to support the inference that h^2 is equal to .80. Jencks argues, on the basis of his analysis of these data, that they imply h^2 is approximately .50. The principle reason for the difference in h^2 estimates derived by Jencks from these data deals with the assumed influences of covariance between genotypes and environments. Jencks's elaborate path analysis does not assume substantial relationship between the environments of the homes in which the separated twins were reared, nor does he assume a substantial influence of the shared prenatal environments of the MZ twin pairs. However, Jencks assumes that the covariance of genetic and environmental characteristics is reduced when a child is raised by parents other than his natural ones. This assumption implies that the variance of separated twins will be less than the variance for twins raised by their natural parents and, accordingly, that the genotypes of twins will account for a greater portion of the total variance in separated pairs than in the population in general. Despite differences in the estimates of h^2 derived from separated twin pairs, both Jencks and Jensen and Herrnstein would agree that such data do support the inference that genetic variables influence test scores.

KAMIN'S CRITIQUE OF SEPARATED TWIN STUDIES

Burt's study has the largest sample (53 pairs) and provides the strongest evidence for the inheritance of intelligence test scores. Burt reports a correlation .86, and, in addition, Burt asserts that the environments in which the twin pairs were raised were substantially uncorrelated. Kamin has three fundamental criticisms of Burt's research. First, there is a notorious absence of information about procedural detail in Burt's reports of his research. No descriptions exist with respect to the procedures used to locate the separated twin pairs. More astounding, it is impossible to reconstruct from Burt's description of his research the precise tests used. On occasion, his descriptions imply that he used an English version of the Binet test. On other occasions, in response to criticisms, he emphatically states that he did not use a Binet but performance tests. No information is presented indicating the exact performance test used. Burt also reports data for group tests without indication of the test used. His most dramatic results, however, are based on what are called final assessments of intelligence. These assessments appear to involve retests of questionable results. The questionable results appear, in part, to be determined by submitting test results to teachers and retesting those children for whom the teacher believed the tests did not adequately reflect the students' ability. However, the exact procedure followed in the reassessment, the tests used for the purpose, and the number of children reassessed are simply not reported. In summary, procedural details for Burt's work are virtually absent.

Second, some of the data reported by Burt are strangely consistent despite relatively large changes in sample size. For example, in 1955, Burt reported a correlation of .771 for 21 MZ twins reared apart on a group test of intelligence. In 1966, he reported the results of his extended sample of 53 separated DZ twin pairs on a group test of intelligence (presumably the same as that used earlier) and again reported a correlation of .771. There are several other equally strange consistencies in Burt's correlations.

Third, analyses of some of Burt's data by Kamin suggest unusual internal inconsistencies. For example, a 1959 paper reporting on 42 separated twin pairs indicated that at least four children of professional parents had been reared in "orphanages." The 1966 Burt report on 53 pairs of separated twins including the previously reported 42 pairs as a subset of the 53, states in direct contradiction that only two children of such parents had been reared in "residential institutions." A number of other such inconsistencies occur in Burt's several reports of his data. Kamin states his final assessment of Burt's data as follows:

> The first serious Burt paper on twins was the work of a 72-year-old gentleman with strong opinions. The majority of the cases described in his final report were collected between that time and his 83rd year. The revised lists of I.Q.'s and social class ratings were being dispatched around the globe in his 88th year.
>
> The conclusion seems not to require further documentation, which exists in abundance. The absence of procedural description in Burt's reports vitiates their scientific utility. The frequent arithmetical inconsistencies and mutually contradictory descriptions cast doubt upon the entire body of his later work. The marvelous consistency of his data supporting the hereditarian position often taxes credibility; and on analysis, the data are found to contain implausible effects consistent with an effort to prove the hereditarian case. The conclusion cannot be avoided: The numbers left behind by Professor Burt are simply not worthy of our current scientific attention [1974, p. 47].

It is extremely difficult to evaluate Kamin's critique of Burt's work, or indeed to evaluate the value of Burt's work in light of this critique. Most scientists, perhaps naively, assume that, independent of interpretation, data are inviolate and are usually reported with great care. The inconsistencies and surprising consistencies do, however, cast some doubt on the work. Whether they indicate some or all of the data are to be distrusted is difficult to decide.

There are three other studies reporting correlations between intelligence test scores for separated MZ twins. Shields (1962) reported a correlation of .77 for 37 separated MZ pairs, using a combined score on a vocabulary test and a nonverbal test (Dominoes) which required extensive verbal instructions from the test administrator. The results for the same tests given to 34 pairs of MZ twins reared together was .76. Kamin has a number of criticisms of these data. Shields makes clear that virtually all of his separated twin pairs were raised in homes that did not differ substantially in social class background. Thus, these twins were not reared in uncorrelated environments. Furthermore, many of the twin pairs were raised in homes of relatives and many had extensive contact with each other. As a matter of fact, few were separated from birth.

A second major criticism is that most of the separated twins were tested by Shields himself with clear knowledge on his part of the performance on each subject's twin. For the five twin pairs not jointly tested by Shields (who were also the most geographically separated) the intraclass correlation was .11 and for the remaining twin pairs tested by Shields the intraclass correlation was .84. These findings suggest the possibility that the similarity of test score performance among the separated MZ twins was in part due to test bias introduced by the examiner.

The only American study of separated MZ twins is that of Newman, Freeman, and Holzinger (1937). Kamin has two principal criticisms of this study. First, the sample of separated MZ twins was radically different from the sample of MZ twins reared together. The latter were school children attending Chicago schools. The separated MZ twins were volunteers who responded to nationwide advertisements and who varied in age from 11 to 59. The majority of the sample of MZ twins who were reared apart were adults. In addition, volunteers among the separated twin pairs were rejected for study if they indicated that they were in any way different from each other, e.g., different in temperament. This was done in order to exclude possible DZ pairs. However, this exclusion might have excluded psychologically dissimilar MZ pairs who might have shown large intrapair differences in intelligence test score.

Perhaps Kamin's most fundamental criticism of these data stems from the fact that there is a strong confounding of age in this study. If intelligence test scores vary with age and if we have separated MZ twin pairs who differ substantially in age at time of test we will find similarity of intelligence among separated MZ twin pairs due solely to the fact that they are, of necessity, the same age. Since we know that published age standardizations of intelligence tests are likely to be wrong due to the cohort effect—i.e., due to the fact that intelligence has increased over time—we would expect that age and intelligence test scores would be negatively correlated (assuming that quotients have been derived from an age standardization sample that was obtained cross sectionally at a single time).

In the Newman, Freeman, and Holzinger study the correlation between age and Intelligence Quotient was −.22. The influence of age can be removed from the correlation by use of partial correlation and this correction yields a correlation for separated twin pairs of .65. Kamin objects to this procedure. He asserts:

> The use of partial correlation assumes that the age–I.Q. relation is linear across the entire range of ages. The procedure also assumes, when both sexes are represented in a sample, that the age–I.Q. relation is the same for each sex. We note that, for the entire sample of 38 individuals, the age–I.Q. correlation was only −.22. For the 14 males, however, it was −.78; for the 24 females, it was −.11. The correlations differed significantly between the two sexes. This apparent sex effect, however, is itself very largely confounded with age. The male pairs averages 20.3 years of age (ranging from 13.5 to 27), while the female pairs averaged 29.8 years (ranging from 11.4 to 59.2). The age variance of the female pairs was significantly greater than that of the males. With the 1916 Stanford-Binet, it is entirely possible that the regressions of I.Q. on age vary, for different sectors of the age range, in sign as well as in slope. Though the sample size is embarrassingly small, it seems prudent to examine the age–I.Q. correlation among females in different age ranges [1974, p. 59].

Kamin attempts to correct for age in this sample by using what he calls a pseudopairing procedure. The procedure involves listing of twin pairs in order of their age. Each twin pair is matched with the twin pair adjacent to it in age. Each of these four scores is pseudopaired with each of the other scores omitting the pairing of scores from actual twins. This procedure permits one to obtain indices of similarity (the intraclass correlation) among biologically unrelated pseudotwin pairs who are similar in age. Kamin reports the results of this analysis by beginning with the seven male pairs (on the assumption that the age–I.Q. confound for males is different from females'). The pseudopairing intraclass correlation is .67 for these seven males (compared to the true correlation for this group of .58). This implies that the correlation is entirely accounted for by the "age confound." When the three female pairs who fall within the male age range are added to the pseudopairing analysis, the correlation becomes .47 (compared to a true correlation of .65). For the seven pairs of females older than the male sample, the observed IQ correlation is .48 and the pseudopairing produces a value of .06. Kamin omits an analysis of the two younger female pairs. This analysis leads Kamin to the conclusion that the correlation between separated MZ twins is substantially attributable to age rather than to genetic similarity.

Kamin's pseudopairing analysis is based on the assumption that the relationship between age and intelligence is different between the sexes and at different portions of the age range. Actually, there is relatively little evidence for a consistent effect of this sort other than the negative correlation of age and IQ. Also, we are aware of no evidence indicating that the inadequacies of age standardization that exist in tests are different for different sexes. Accordingly, Kamin's decision to apply separate analyses to control for age in different samples is based on the demonstration that there is a significant sex difference in the relationship between age and IQ in this sample of 19. Actually, Kamin's statistical test for this difference is not legitimate. The appropriate statistical test in our judgment should not compare 24 females with 14 males. One cannot consider 12 MZ twins as forming 24 independently sampled individuals. A more appropriate comparison would involve the comparison between 12 female pairs (taking the average score for a pair) with 7 male pairs. For these sample sizes, the appropriate standard error of the difference between the correlations for samples would be approximately .6. In order to assert that there is a significant difference between the sexes in correlations in a sample of this size one would have to find, for example, that correlation with age in one sex approaches unity and is in the opposite direction in the other sex. Thus Kamin's analysis appears to be based on an untenable assumption. The pseudopairing procedure should, in our judgment, have been applied to the total sample. This procedure would have in all probability produced a trivial correction for age not fundamentally different from that produced by the partial correlation technique. Alternatively, Kamin could have applied the pseudopairing procedure to the 12 pairs of female MZ twins. It seems somewhat arbitrary to divide a sample of 12 pairs into 3 groups (one "juvenile"), one within the age range of the male sample, and one aged. The pseudocorrelation procedure for the females alone again would, in all probability, have indicated that the relationship between female MZ twins was not substantially influenced by age. Kamin's pseudopairing procedure seems more capricious and arbitrary than a correction based on the partial correlation technique. It should also be noted that the partial correlation technique is not terribly dependent on a linear assumption. For all practical purposes, the technique is justified as long as there is a monotonic relationship with age. We suspect that the pseudopairing technique cannot be used to explain away the similarity of the separated MZ twin pairs. Of course, this does not vitiate Kamin's critique of this study which is based on the unusual and nonrepresentative character of the separated twin pairs.

The last study of separated MZ pairs we shall discuss is a Danish study performed by Juel-Nielsen (1965). The study had only 12 twin pairs—9 male and 3 female. A nonstandardized version of the WAIS was used. Juel-Nielsen reports a correlation among separated twin pairs of .62. Again an age confound exists. The correlation for the 18 females with age was a statistically significant value of .61. And, the correlation for the 6 males was a statistically significant −.82. Kamin reports that, for the 9 female twin pairs, the intraclass correlation was .59 and a pseudopairing procedure yielded a correlation of .59. This implies that the relationship in intelligence test scores among these twin pairs is not due to genetic identity but rather to age identity.

Again, Kamin's analysis depends on the legitimacy of controlling for age separately by sex in such small samples. First, use of the appropriate sample sizes 9 and 3 twin pairs for female and male samples, respectively, would imply that the appropriate values of correlation required to reject the null hypothesis would be .666 and .997 respectively. And, no values of r for each of these samples could be judged significantly different by statistical test. Again, correction for age in which the sample is pooled even using the pseudopairing procedure would suggest that the age-IQ confounding is not as major a determinant of the relationship between the separated MZ twins as is implied by Kamin.

Quite apart from the question of age confound, some of the twin pairs in the Juel-Nielsen study clearly were reared in similar environments and this would account for some of their similarity in intelligence test score.

What can be concluded about the inheritance of intelligence test score, on the basis of the separated twin studies? We do not find Kamin's criticism of these studies with respect to age confounding completely convincing. On the other hand, the fact that separated twins are rarely (except in Burt's study) raised in uncorrelated environments and that the samples of such twins are probably not very representative indicates at very least that there is some question about their value as absolutely critical data. On the other hand, we do not believe that the data can be dismissed. The fact that separated MZ twins appear to resemble each other in test score despite some divergence in environment tends to support the inference that intelligence test scores are at least partially determined by genetic characteristics.

Conclusion: Genetics and Intelligence

We are ready to venture some overall conclusions about the question of possible genetic determinants of intelligence test scores. We believe that Kamin's attempt to look at the evidence for the genetic influence on intelligence test scores from the viewpoint of a confirmed environmentalist has been extremely useful. He has succeeded in casting a penumbra of uncertainty around what has heretofore been a literature frequently interpreted (or, for that matter, misinterpreted and overinflated in its import by confirmed hereditarians) as conclusively demonstrating a genetic influence on intelligence test scores. Indeed, in some of the more biased or extreme summaries of this literature, genetic influences are considered to be so preeminent that the contribution of environment to test score is nil. Also, Kamin's analysis has convinced us that some of the extremely elegant applications of population genetic models such as Jencks's use of path analyses and Jinks and Fulker's (1970) models are probably premature in that the data used in these models are not that good.[2] However, for the reasons we indicate earlier we do not believe that Kamin has succeeded in his goal of indicating that the belief in genetic influences on intelligence test score is one without foundation at the present time. On the contrary, we believe that there is a reasonable degree of evidence for the proposition that intelligence test scores are influenced to some degree by genotypes. Furthermore, the available data suggest that the degree of influence may be somewhat higher than the value assumed by Jencks.

Biological Environment and Intelligence

The influence of the environment on intelligence test scores is not in principle limited to the characteristic cognitive socialization experiences of an individual. There is a great deal of speculation and research related to the potential influence of the biological environment on intellectual development. There is a relatively wide range of potential biological influences on test scores. Many of these are associated with gestation and birth. One impetus to the search for biological influences on intellectual development derives from the notion of the "continuum of reproductive casualty" (Pasamanick & Knobloch, 1966). According to this concept, a number of children who survive and who may not show any clear overt signs of neurological impairment have suffered some degree of neurological impairment during gestation and birth. Such impairments are assumed to be associated with pregnancy and birth complications and indeed are

[2] Jencks apparently would not totally disagree with this since he frequently mentions that the data base is rather more fragile than many of the secondary source summaries indicate.

assumed to be caused by such complications. These impairments, in turn, are assumed to be associated with subsequent maladjustment including the potential for somewhat retarded intellectual development. The concept of a continuum of reproductive casualty has influenced research on such topics as the intellectual consequences of anoxia and of low birth weight.

Interest in the potential influence of biological variables on intelligence has also been spurred by the publication of a book by Birch and Gussow (1970) entitled *Disadvantaged Children: Health, Nutrition and School Failure.* In this book Birch and Gussow summarize data that indicate that children growing up in poverty are likely to be exposed to poorer medical care and are likely to have poorer diets. They argue that these deficiencies in the biological nurturance of children are related to subsequent school failure.

Our discussion of research on the effects of the biological environment on intelligence tests will include a discussion of research relating such variables as birth weight and prematurity, anoxia, and other complications of pregnancy and delivery, and nutrition to intelligence.

Continuum of Reproductive Casualty and Intelligence

A number of studies have dealt with the relationship between low birth weight and subsequent intellectual development. A number of cautions should be exercised in interpreting the results of such studies. First, low birth weight is an ambiguous variable in that the class of children of low birth weight includes both children who are born prematurely and children who are born at or near to term but who are born small for their gestational age. Second, as with any naturally occurring event, comparisons between children born with low birth weight and children born with higher birth weights do not permit us to infer with certainty that the differences, if any, which exist between these groups are attributable to birth weight per se or are attributable to any of a large number of respects in which these children may differ from normal-birth-weight children. The variables associated with low birth weight might include a variety of social and biological characteristics of parents and even might include socialization practices. It is conceivable that parents of low-birth-weight children tend to subtly treat their children differently in ways that might impede their intellectual development. We shall see that the relationship between variables such as birth weight and intelligence test scores is relatively small. Therefore, it is difficult to know whether birth weight itself is an important influence on intellectual development or whether its influence is indirect and mediated by other variables with which it is correlated.

The most dramatic evidence of intellectual declines associated with low birth weight comes from studies dealing with children whose birth weight was extremely low. For example Drillien (1964) found developmental quotients averaging 80.2 in a sample of 40 4-year-old children with birth weights under 3 lb 9 oz compared to quotients ranging between 97 and 107 in children of higher birth weight from the same socioeconomic background. The decrement in intelligence test scores associated with low birth weight has been found to persist for a considerable number of years. Weiner (1968) has reported the results of intelligence tests given to 848 children at age 12 to 13 in an original sample including 500 low-birth-weight children and a group of individually matched controls with birth weight in excess of 2500 gm (see Knobloch *et al.*, 1956, for a fuller description of the study). He found a correlation of .14 between birth weight and IQ obtained from school records for this sample.

The data reported by Drillien and by Weiner are based on studies in which a group of low birth weight children are compared to a group of non-low-birth-weight children. Birth weight is a continuous variable and its relationship to intelligence test scores can be assessed in a nonselected sample. The most important data of this kind come from the collaborative perinatal project (Broman, Nichols, & Kennedy, 1975). This study included a sample of 53,043 women giving birth in a variety of hospitals in the United States. A large body of data was collected on the social background of the parents as well as information about pregnancy, birth, and delivery. The children resulting from these pregnancies were studied in a longitudinal design which included the administration of a Stanford-Binet IQ at age 4. Figure 5.1 presents data indicating the relationship between birth weight and IQ at age 4 for four groups of children—black and white male and female children. The data in Figure 5.1 indicate that IQ at age 4 increases as birth weight increases. The data also provide evidence of a curvilinear trend with the largest differences occurring between the lowest birth weight children and those weighing between 2500 and 3000 grams. Above that level there appears to be relatively little difference and there is even the suggestion of a downturn in IQ for those children with the highest birth weight. The correlation between birth weight and IQ at age 4 was .07 for the white sample, .11 in the black sample, and .17 in the total sample.

The collaborative perinatal project was designed to provide data

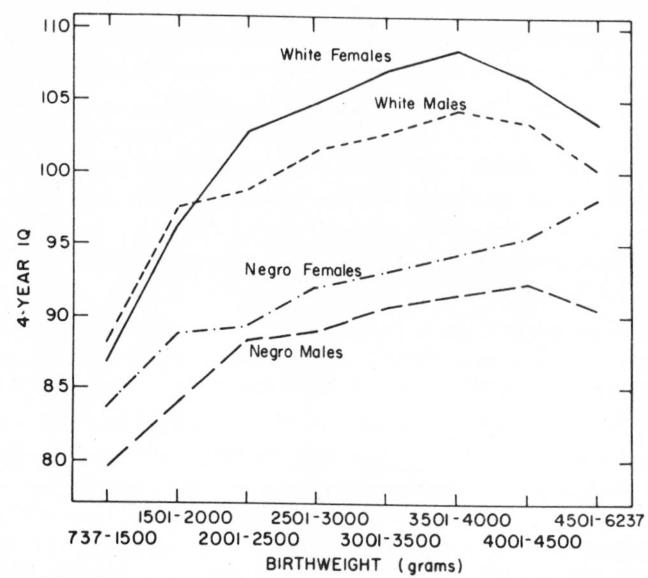

Figure 5.1. *Four-year IQ by birthweight, race, and sex. [From Broman, S.H., Nichols, P.L., and Kennedy, W.A.,* Preschool IQ: Prenatal and early developmental correlates. *Hillside, New Jersey: Lawrence Erlbaum Associates 1975. Reproduced by permission.]*

about a large number of biological and social variables that might relate to subsequent intellectual development. Table 5.4 presents the results of correlations between 132 variables relating to the social and medical history of the family and to biological factors relating to pregnancy, birth, and delivery. A number of important conclusions can be derived from an examination of the data in Table 5.4 and from supplemental information provided by Broman, Nichols, and Kennedy. First, it is apparent, in line with previous findings, that racial characteristics and socioeconomic indices relate to intellectual development (see variables 1 to 7 in the table). The set of variables related to family medical history—variables 6 to 32—are simply not important determiners of IQ in the population as a whole. In interpreting these findings it should be noted that they reflect in part the frequency of occurrence of a particular event. For example, the mean IQ of children whose mothers are retarded (variable 8) may be dramatically lower than the mean IQ of children whose mothers are not classified as retarded. However, a relatively small percentage of children in a sample roughly representative of the population will have retarded mothers. The correlation of −.06 for this variable indicates that in predicting the IQ of all children in a representative sample, knowledge of the mother's status with respect to retardation is not a particularly useful item of information. Notwithstanding the possibility that in the relatively small number of cases in which a child has a mother classified as retarded, it is possible from knowledge of this fact alone to predict for such children a probable low IQ score. Excluding variables 1 to 7, only 14 of the 125 other variables considered have a correlation with IQ of .11 or greater. And, even these variables do not unambiguously reflect the influence of biological factors on intellectual development. In fact in some cases an inference of biological influences can be reasonably eliminated.

Let us consider each of these correlations. Variable 33 relates to the familiar finding that birth order is related to intelligence. This variable is specifically discussed in Chapter 6 and we indicate there that the influence of birth order may be attributable to patterns of social interaction in the family. Variable 52, number of prenatal visits, is in part related to socioeconomic status. However, even with socioeconomic status controlled, the variable has a significant relationship to IQ at age 4. However, the relationship may relate to socialization practices. Mothers who visit their physician frequently during pregnancy may provide their children with a more appropriate learning environment. Alternatively, such mothers may be, as a group, more intelligent than mothers who visit their physicians infrequently. Variables 53, 54, and 55 on the surface appear to indicate direct relationships between biological conditions during pregnancy and intelligence at age 4. However, each of these variables is related to race and socioeconomic status. And, the relationship between them and intelligence substantially decreases when socioeconomic status and race are controlled. Such a result is inherently ambiguous in its appropriate interpretation. It may be interpreted as indicating that the relationship between these biological variables and intelligence is indirect and is related to the fact that individuals with different social status may have different genotypes and may also provide different cognitive socialization experiences for their children. On this interpretation, changes in levels of anemia during pregnancy would not affect intelligence. Alternatively, it might be argued that one reason that there are differences in intelligence between groups with different social status is that these groups differ in the probability of providing optimal biological environments during pregnancy. Variable 56, gestation at registration, is correlated with variable 52, number of prenatal visits. That is, mothers who began prenatal care early were likely to have children with higher IQ at age 4. Again, this correlation is partially due to socioeconomic

TABLE 5.4
Relationship between prenatal and Neonatal Characteristics and Intelligence at Age 4[a]

	White	Black	Total
Population characteristics			
1. Race—White versus black			−.40
Family characteristics			
2. Socioeconomic index	.38	.24	.43
3. Mother's SRA score	.28	.18	.36
4. Mother's education	.38	.21	.35
5. Mother's marital status			
Never married versus married	.07	.08	.19
Previously married versus married	.11	—	.10
All unmarried versus married	.12	.07	.19
6. Mother's employment status	.15	.07	.12
7. Presence of father in the home	.13	.06	.19
Family History			
8. Retardation in mother	−.06	−.05	−.06
9. Congenital malformations in mother	—	—	.03
10. Mental illness in mother	—	—	—
11. Motor defects in mother	—	—	—
12. Seizures in mother	—	—	—
13. Sensory defects in mother	—	—	—
14. Mental illness in father	−.04	—	—
15. Sensory defects in father	−.04	—	—
16. Congenital malformations in father	—	—	.04
17. Diabetes in father	—	—	—
18. Retardation in father	—	—	−.03
19. Motor defects in father	—	—	.03
20. Seizures in father	—	—	—
21. Retardation in full siblings	−.06	—	—
22. Congenital malformations in full siblings	—	—	.03
23. Motor defects in full siblings	—	—	—
24. Rh incompatibility in full siblings	—	—	.03
25. Seizures in full siblings	—	—	—
26. Sensory defects in full siblings	—	—	—
27. Congenital malformations in half siblings	—	—	—
28. Retardation in half siblings	—	—	—
29. Motor defects in half siblings	—	—	—
30. Rh incompatibility in half siblings	—	—	—
31. Seizures in half siblings	—	—	—
32. Sensory defects in half siblings	—	—	—
Maternal Characteristics			
33. Parity	−.14	—	−.11

Continued

[a]From Broman, S.H., Nichols, P.L., and Kennedy, W.A. *Preschool IQ: Prenatal and early developmental correlates.* Hillside, New Jersey: Lawrence Erlbaum Associates, 1975. Reproduced by permission.

TABLE 5.4—*Continued*

	White	Black	Total
34. Gravidity	– 14	—	–.10
35. Age	.03	.10	.10
36. Birthweight of last liveborn child	—	.04	.09
37. X-ray exposure history			
None vs. abdomino-pelvic	.10	.16	.19
None vs. other areas	.06	.07	.08
Scale (0–2)[a]	—	.06	.08
38. Smoking history	—	.06	.08
39. Menstrual cycle interval			
Normal (21–35 days) vs. < 21 days	—	—	—
Normal vs. > 35 days	.05	—	.07
Normal vs. < 21 days and > 35 days	—	—	.03
Normal vs. "irregular"	—	—	—
Normal vs. all unusual intervals	—	—	.04
40. Height	.09	.03	.06
41. Confining illnesses	.04	—	.06
42. Interval since last pregnancy	.05	.07	.06
43. Sterility investigation	.04	.04	.06
44. Age at menarche	–.05	–.03	–.04
45. Outcome of last pregnancy			
Surviving child vs. fetal death	—	.04	—
Surviving child vs. neonatal death	—	—	—
Surviving child vs. fetal death or neonatal death	—	—	—
46. Prepregnant weight	—	.05	—
47. Rh blood type—positive vs. negative	—	—	.07
48. ABO blood type			
All others vs. A	—	—	.06
All others vs. B	—	—	–.06
All others vs. AB	—	—	—
All others vs. O	—	—	—
49. Number of prior stillbirths	—	—	—
50. Number of prior stillbirths and neonatal deaths	—	—	–.04
51. Length of time to become pregnant	—	—	—
Prenatal period			
52. Number of prenatal visits	.21	.13	.25
53. Lowest hematocrit of mother during pregnancy	.07	.07	.22
54. Lowest hemoglobin of mother during pregnancy	.12	—	.21
55. Anemia during pregnancy	–.08	–.03	–.16
56. Gestation at registration	–.20	–.10	–.19
57. Duration of pregnancy	.04	.07	.12
58. Edema during pregnancy	—	.05	.10
59. KUB infection during pregnancy	–.06	—	–.09
60. Cigarettes smoked per day during pregnancy	–.07	.04	.07
61. Hypertensive blood pressures during pregnancy	.03	—	.04
62. Hospitalization during early pregnancy	–.07	—	–.03
63. Convulsions during pregnancy	—	–.03	–.02

Continued

TABLE 5.4—*Continued*

	White	Black	Total
64. Maximum weight gain during pregnancy	−.03	—	—
65. Pelvic summation (inlet)			
Adequate versus borderline	—	—	—
Adequate versus contracted	—	—	−.02
Adequate versus borderline or contracted	—	—	−.03
Scale (0–2)[b]	—	—	−.03
66. Acute or chronic asthma during pregnancy	—	—	—
67. Acute or chronic glomerulonephritis during pregnancy	—	—	—
68. Diabetes during pregnancy	—	—	—
69. Fever during pregnancy	—	—	.02
70. Jaundice during pregnancy	—	—	—
71. Urine specimens with albumin 2+ or more	—	—	−.03
72. Urine specimens with glucose 2+ or more	—	—	—
73. Organic heart disease during pregnancy	—	—	—
74. Rheumatic fever during pregnancy	—	—	—
75. Syphilis during pregnancy	—	—	−.05
76. Vaginal bleeding during pregnancy	—	—	.03
77. Vomiting during pregnancy	—	—	—
Labor and delivery			
78. Use of forceps during delivery			
None versus class I	.13	—	.14
None versus class II	.12	—	.14
None versus class III	.10	—	.14
None versus class IV	—	—	—
Scale (0–4)[b]	.10	—	.15
79. Placental weight	—	.09	.10
80. Duration of first stage of labor	—	—	—
81. Duration of second stage of labor	.06	—	.10
82. Augmentation of labor			
None vs. oxytocic	.06	—	.05
None vs. Amniotomy	—	—	—
None vs. combination of methods	—	—	—
None vs. all methods	.05	—	—
83. Lowest fetal heart rate during first stage of labor	—	—	−.09
84. Highest fetal heart rate during first stage of labor	.05	—	.03
85. Lowest fetal heart rate during second stage of labor	—	—	−.09
86. Highest fetal heart rate during second stage of labor	.06	—	—
87. Presence of fetal heart sounds at admission for delivery	.03	—	.02
88. Vaginal bleeding at admission for delivery	—	—	—
89. Meconium staining	—	—	—
90. Induction of labor			
None vs. oxytocic	—	—	.03
None vs. amniotomy	—	—	.06
None vs. mechanical	—	—	—

Continued

TABLE 5.4—*Continued*

	White	Black	Total
None vs. combination of methods	—	—	.04
None vs. all methods	—	—	.06
91. Arrested progress of labor			
Phase			
None vs. latent phase	—	—	.02
None vs. active phase	—	—	—
None vs. second phase	—	—	.02
None vs. two or more phases	—	—	—
None vs. all phases	—	—	.03
Probable cause			
None vs. disproportion	—	—	—
None vs. malpresentation	—	—	.04
None vs. uterine dysfunction	—	—	—
None vs. miscellaneous causes	—	—	—
None vs. combination of causes	—	—	—
None vs. all causes	—	—	—
92. Ruptured membrane at or after onset of labor	—	—	−.03
93. Interval between rupture and onset of labor	—	—	—
94. Uterine dysfunction	—	—	.03
95. Caesarean section after onset of labor	—	—	—
96. Primary indication for Caesarean section			
Previous section versus older primipara	—	—	—
Previous section versus toxemia	—	—	—
Previous section versus diabetes mellitus	—	—	—
Previous section versus failed pelvic procedure	—	—	—
Previous section versus malpresentation	—	—	—
Previous section versus uterine dysfunction	—	—	—
Previous section versus fetal distress	—	—	—
Previous section versus prolapsed cord	—	—	—
Previous section versus placenta previa	—	—	—
Previous section versus abruptio placenta	—	—	—
Previous section versus other indications	—	—	—
Previous section versus all other indications	—	—	—
97. Forceps delivery of head in Caesarean section	—	—	—
98. Artificial rupture of membrane at delivery	—	—	—
99. Delivery complicated by polyhydramnios	—	—	—
100. Type of delivery			
Vertex versus breech	—	—	—
Vertex versus Caesarean section	—	—	—
Vertex versus all nonvertex	—	—	—
101. Presentation at delivery			
Occipito-anterior (OA) versus occipito-posterior (OP)	—	—	—
OA versus breech	—	—	—
OA versus other (face, brow, transverse)	—	—	—

Continued

TABLE 5.4—*Continued*

	White	Black	Total
OA versus OP, breech and other	—	—	—
OP versus breech	—	—	—
OP versus other	—	—	—
OP versus OA, breech and other	—	—	—
Breech versus other	—	—	—
Breech versus OA, OP and other	—	—	—
Other versus OA, OP and other	—	—	—
102. Vacuum extraction in vertex deliveries	—	—	—
103. Difficulty of forceps usage, vertex deliveries	—	—	—
None versus moderate	—	—	—
None versus severe	—	—	—
None versus failure	—	—	—
None versus all degrees	—	—	—
Scale (0–3)[a]	—	—	—
104. Extraction in breech delivery			
Partial extraction			
No extraction vs. average difficulty	—	—	—
No extraction vs. difficulty	—	—	—
No extraction vs. very difficult	—	—	—
No extraction vs. all partial extraction	—	—	—
Scale (1–3)	—	—	—
Total extraction			
No extraction vs. average difficulty	—	—	.38
No extraction vs. difficulty	—	—	—
No extraction vs. very difficult	—	—	—
No extraction vs. all total extraction	—	—	.37
Scale (1–3)[a]	—	—	—
105. Cord complications			
None vs. true cord knot	—	—	—
None vs. cord around neck	—	—	.04
None vs. cord around body	—	—	.02
None vs. prolapsed cord	—	—	—
None vs. all complications	—	—	.04
106. One umbilical artery	—	—	—
107. Placental complications			
Abruptio placenta			
None vs. partial	—	—	—
None vs. complete	—	—	—
None vs. partial or complete	—	—	—
Scale (0–2)	—	—	—
Placenta previa			
None vs. total implant	—	—	—

Continued

TABLE 5.4—*Continued*

	White	Black	Total
None vs. partial	—	—	—
None vs. marginal	—	—	—
None vs. low implant	—	—	—
None vs. all placenta previa	—	—	—
Marginal sinus rupture	—	—	—
None vs. all placental complications	—	—	—
Neonatal Period			
108. Birthweight	.07	.11	.17
109. Length at birth	.06	.11	.15
110. Head circumference at birth	.08	.10	.15
111. Sex	.11	.08	.08
112. Neonatal brain abnormality	−.09	−.04	−.06
113. Highest neonatal serum bilirubin	−.03	−.07	−.05
114. Neonatal respiratory distress	−.05	−.04	−.05
115. Neonatal direct Coombs' Test	.03	—	.05
116. Primary apnea	−.04	—	−.04
117. Simgle neonatal apneic episode	−.03	—	—
118. Multiple neonatal apneic episodes	—	—	—
119. Resuscitation during first 5 minutes of life	—	—	—
120. Resuscitation after first 5 minutes of life	—	—	—
121. One minute Apgar score	.04	—	−.02
122. Five minute Apgar score	—	—	−.04
123. Lowest neonatal hematocrit	−.06	—	—
124. Lowest neonatal hemoglobin	—	—	−.06
125. Dysmaturity			
None versus equivocal	—	—	—
None versus Stage I	—	—	—
None versus Stage II	—	—	—
None versus Stage III	—	—	—
None versus all stage	—	—	.03
Scale (0–4)[b]	—	—	.03
126. Neonatal cephalohematoma	—	—	.02
127. Neonatal spinal cord abnormality	—	—	—
128. Neonatal peripheral or cranial nerve abnormality	—	—	—
129. Neonatal fractured skull	—	—	—
130. Neonatal intracranial hemorrhage	—	—	—
131. Neonatal central nervous system infection	—	—	—
132. Neonatal clinical erythroblastosis			
None versus cases without transfusion	—	—	—
None versus cases with transfusion	—	—	—
None versus all cases	—	—	—
Scale (0–2)[b]	—	—	—

status. Within various socioeconomic groups the correlation drops below .05. Variable 58, edema during pregnancy, is correlated with race and within each race and within socioeconomic groupings it is not related to IQ at age 4. Variable 78, the use of forceps during delivery, indicates, unexpectedly, that the use of forceps is associated with higher IQs at age 4. However, the use of forceps is positively associated with socioeconomic status and inversely with parity (children born later are less likely to have forceps deliveries). When these two variables are controlled, the relationship between use of forceps and intelligence is negligible. The positive correlation for variable 104, extraction in breech delivery, is difficult to interpret. The finding indicates that among the set of children with breech deliveries those children delivered without extraction had substantially higher IQs at age 4 than did those children who were delivered using total extraction procedures. However, the correlation was not present in either the white or black sample considered by itself—suggesting that this variable may be confounded with race. Also, the correlation is the only significant correlation of any magnitude that is not discussed in the text. Presumably, Broman, Nichols, and Kennedy consider the correlation unimportant or in some way artifactual. Variables 108, 109, and 110 relate to the size of the neonate at birth and indicate that neonates who are larger at birth are likely to have higher IQ at age 4. These correlations, though relatively low, are about the only measurable biological variables associated with gestation, birth, and delivery that have some influence on IQ at age 4 within different racial and socioeconomic groups.

Our review of the variables investigated in the collaborative perinatal project indicates that apart from socioeconomic status and race and with the possible exception of neonatal size, there is little indication in these data that information related to the nature of gestation, delivery, and birth is of much value in predicting subsequent intelligence. Within the population as a whole, information about social and racial status is of far more importance in predicting subsequent intellectual performance. Further, these data suggest that the continuum of reproductive casualty is not of major significance in accounting for individual differences in intelligence. This conclusion can be buttressed by an analysis of the development of stepwise regression equations to predict IQ at age 4 using as predictor variables the set of 132 variables we have considered. Broman, Nichols, and Kennedy present four such stepwise regressions for black and white male and female children. Table 5.5 presents an analysis of the relation between these variables and IQ among white male children.

TABLE 5.5
Stepwise Regression Analysis in Four Samples Relating Prenatal and Neonatal Variables to IQ at Age 4[a]

Sample	Contribution of mother's education and socio-economic status (r)	Final contribution considering all other significant variables (r)
White male	.42	.46
White female	.42	.47
Black male	.25	.30
Black female	.28	.34

[a]Based on Broman, Nichols, and Kennedy (1975).

Virtually all of the predictive variance in IQ was accounted for by mother's education and socioeconomic background in each of the four analyses. The results of the collaborative perinatal project represent the most systematic and largest scale analysis of variables associated with the continuum of reproductive casualty that relates to intelligence. The results obtained, however, are not anomalous. For example, Werner *et al.* (1971) have reported on the results of a similar study performed on the island of Kauai relating social background, prenatal, and neonatal characteristics to intellectual and academic performance at age 10. They too report that social characteristics are far more predictive of subsequent development than are variables related to the process of gestation and birth.

One variable that may be associated with prenatal events has been found to be associated with subsequent intelligence. Waldrop and Halverson (1971) have devised a measure of minor physical anomalies present at birth and at other ages as well. The list of anomalies are those that have been found to be associated with Down's syndrome and are somewhat more likely to be present in children with major neurological defects. The list of defects is presented in Table 5.6. The score for the presence of these anomalies has been found to be related to hyperactivity, especially among males, in elementary school children.

In one of their studies, Waldrop and Halverson report a correlation of −.34 for $7\frac{1}{2}$-year-old males between the anomalies scores and IQ and a correlation of −.46 for the females in their study. These results were replicated by Tignor in a study of a group of suburban school children. She found a correlation of −.21 between IQ and the anomalies scores in her sample. These studies provide evidence of a

TABLE 5.6
List of Anomalies[a]

Anomaly	Weight
Head	
Fine electric hair:	
Very fine hair that will not comb down	2
Fine hair that is soon awry after combing	1
Two or more hair whorls	0
Head circumference outside normal range:	
$> 1.5\ \sigma$	2
$> 1.0\ \sigma \leqslant 1.5\ \sigma$	1
Eyes	
Epicanthus:	
Where upper and lower lids join the nose, point of union is:	
Deeply covered	2
Partly covered	1
Hyperteliorism:	
Approximate distance between tear ducts:	
$> 1.5\ \sigma$	2
$> 1.0\ \sigma \leqslant 1.5\ \sigma$	1
Ears	
Low-seated ears:	
Point where ear joins the head not in line with corner of eye and nose bridge:	
Lower by $>$ .5 cm	2
Lower by $\leqslant$ .5 cm	1
Adherent ear lobes:	
Lower edge of ears extend:	
Upward and back toward crown of head	2
Straight back toward rear of neck	1
Malformed ears	1
Asymmetrical ears	1
Soft and pliable ears	0
Mouth	
High-steepled palate:	
Roof of mouth:	
Definitely steepled	2
Flat and narrow at the top	1
Furrowed tongue (one with deep ridges)	1
Tongue with smooth-rough spots	0

Continued

[a]From Waldrop, M.F. *et al.* Minor physical anomalies and behavior in preschool children. *Child Development*, *39*, Table 1, p. 394. Copyright © 1968 by The Society for Research in Child Development, Inc.

TABLE 5.6—*Continued*

Anomaly	Weight
Hands	
Curved fifth finger:	
Markedly curved inward toward other fingers	2
Slightly curved inward toward other fingers	1
Single transverse palmar crease	1
Feet	
Third toe longer than second:	
Definitely longer than second toe	2
Appears equal in length to second toe	1
Partial syndactylia of two middle toes	1
Big gap between first and second toes	1

relatively large relationship between a biological variable measurable at birth and subsequent intelligence. An understanding of the relationship hinges, in part, on an understanding of the etiology of the anomalies. They may be due to chromosomal aberrations during the first trimester of pregnancy or perhaps to a directly genetic influence.

Nutrition and Intelligence

The influence of nutrition on intelligence has been the subject of increasing attention in recent years. Much of this research has centered on the possibility that malnutrition during the prenatal or early postnatal period may lead to structural damage to the brain and as a result lead to inadequate intellectual development. Birch and Gussow (1970) have explored the possibility that some of the variations in intelligence test scores associated with race and social class may be attributable to differences in nutrition and they have suggested that children growing up in poverty may be less likely to be adequately nourished and that this, in turn, might lead to less adequate intellectual development.

The evidence for the influence of nutrition on intelligence test scores is indirect and confused. Some of it derives from animal studies and is of questionable relevance to humans. Other data come from studies of children growing up in countries other than the United States under conditions of extreme nutritional deprivation leading to clinical manifestations of malnutrition; these data are of questionable relevance to any substantial segment of the U.S. population.

In reviewing some of the literature relating nutrition to mental development, we will ignore animal studies. We shall follow a sequence beginning with studies that provide tangential or indirect evidence for a nutritional influence on intelligence and end with studies that attempt to provide data that are directly relevant.

There is evidence that early nutritional deprivation may affect the development of the brain in humans. For example, Winnick, Brasel, and Rosso (1972) have reported on the results of autopsy investigations of children who died of acute malnutrition during the first or second year of life. Children who died of malnutrition in the first year of life, but not those who died in the second year, were found to have fewer brain cells than would be expected normally. Although this analysis provides suggestive data indicating that there are possible structural changes in the brain associated with early malnutrition, it should be noted that in at least two respects such data are of questionable relevance to the assumption that nutrition is related to intelligence test scores. First, it is not known how such possible changes in the brain might relate to subsequent intelligence test results. Second, children who have died of malnutrition represent such a limited and extreme group that it is difficult to infer anything about the influences of variations in nutrition on intelligence in the normal population from data derived from these children.

More direct information exists about the influence of malnutrition on later intelligence. Several studies have dealt with the intelligence test scores of children who were hospitalized for malnutrition. Stoch and Smythe (1968) studied a group of 20 undernourished South African children. They compared the intelligence test performance of these children with a group of control children whose parents had the same intelligence test scores as the parents of the malnourished children. They found that the malnourished children had mean test scores that were approximately 20 points lower than the mean of control group children. Similar dramatic depression in the intelligence test performance of malnourished children has been reported by Champskam, Srikantia, and Gopalan (1968). They studied the mental development of 19 children treated for kwashiorkor—a disease resulting from protein and calorie malnutrition—in Hyderabad, India. Compared to a control group of children matched for socioeconomic status, these children exhibited a dramatic reduction in a variety of test scores.

These two studies as well as several related ones reviewed by Stein and Kassab (1970) provide only indirect evidence for the influence of nutrition on mental development.

In both of these studies, despite the attempt at matching, children who were nutritionally deprived came from families that, on a variety of social indices, were more deprived than were children in the control group. Children experiencing severe malnutrition are likely to be neglected and abused in other ways and it is not clear that the subsequent depression in intelligence they exhibit is a direct result of their earlier malnutrition or of the social environment they experience. One way to provide some control for differences between family backgrounds of severely malnourished children is to compare the test scores of such children with the test scores of their siblings. Hertzig *et al.* (1972) reported the results of tests administered to 71 boys in Jamaica, West Indies, who had been hospitalized for malnutrition. They found that the malnourished children had WISC IQs that were four points lower than their siblings. Similar results were reported by Birch *et al.* (1971) for a group of children hospitalized for kwashiorkor in Mexico City. The children were tested at least 3 years after their release from the hospital. These children had test scores that were 13 points lower than their sibling controls. On the other hand Evans, Moodie, and Hansen (1971) have reported the results of a study in South Africa in which they failed to find a difference between the test scores of children hospitalized for kwashiorkor early in life and their sibling controls.

These studies provide some evidence for the notion that early malnutrition may retard subsequent intellectual development. Again, however, it should be noted that these studies deal with extreme populations and provide little information with respect to the influence of nutritional variations on intelligence test performance in the population of the United States.

There is one recent study that provides information about the influence of early malnutrition on subsequent intellectual development in a broadly representative population. Stein *et al.* (1975) have published a book dealing with an analysis of the military examinations of 19-year-old Dutch selective service registrants who had been subjected to conditions of famine during 1944 and 1945 in the Netherlands during the German occupation of that country. The famine experiences of the parents of these children were severe leading to increases in mortality up to the first 3 months of life, decreased birth weight, and to a lesser extent, decreases in infant length and head circumference. Approximately 20,000 males were studied whose place of birth indicated that their mothers had been exposed to some degree of malnutrition during the prenatal period. The Ravens test scores of this group were compared to the scores of more than 100,000 male selective service registrants at the same time whose place of birth indicated that their mothers had not been exposed to malnutrition during the prenatal period. There was no difference in the Ravens test score performance between these two groups. The results indicate rather conclusively that severe prenatal malnutrition does not lead to irreversible structural changes in the brain that retard subsequent intellectual development. However, these results refer to individuals who experienced adequate postnatal nutrition. It is possible that individuals who chronically and persistently receive inadequate or only marginally adequate diets would suffer some depression in their intelligence test scores.

The studies of nutrition and mental development we have received have not been experimental studies. Since subjects were not randomly assigned to conditions, the possibility exists that results indicating a nutritional effect might be due to some other variable correlated with nutrition. Such a possibility exists even with respect to studies using sibling controls. In this case the sibling who has suffered severe malnutrition might have been disliked or maltreated in other ways and these experiences, rather than malnutrition, might be responsible for lowering test scores.

Such difficulties can be circumvented by the use of true experimental manipulations. Clearly, an experimental study involving nutritional deprivation cannot be performed. However, one involving nutritional supplementation is feasible. At least two such studies have been conducted in the United States. Harrell, Woodyard, and Gates (1956) have reported the results of a study using vitamin supplements during pregnancy. Among a group of black urban residents of Norfolk, Virginia, they found that the use of vitamin supplements during pregnancy led to a small but significant increase in the Stanford-Binet IQs of children resulting from those pregnancies at age 3 to 4 in comparison to the IQs of a randomly selected group of pregnant women who were given placebos. However, when the experiment was repeated in a group of white Kentucky mountain women no measurable differences in the intelligence of children was found. It can be argued that the women in the rural setting had more adequate access to food. Rush *et al.* (1974) have designed a study involving random assignment of pregnant women to groups receiving nutritional supplementations of protein and calories, calories with little protein, and a control group in a sample of black women in

Harlem. One of us (N.B.) has been involved in the administration of batteries of psychological tests to the children resulting from those pregnancies. Preliminary results are available for 500 children resulting from these pregnancies at age 1. The preliminary results indicate that there are no significant differences among these groups of children on such measures of development as the Bayley tests of infant development and Piaget scales of development as a test of the levels of object permanence reached. However, there were interesting differences among the groups on attentional measures. Children whose mothers had received protein supplements during pregnancy showed more rapid habituation to a repeatedly presented visual stimulus, a greater magnitude of recovery of visual attention (dishabituation) to a change in the stimulus, and, during play, these children showed longer average duration of play with the same toy relative to children whose mothers received caloric supplementation during pregnancy or whose mothers were assigned to the control group. Unfortunately, actual intelligence test data at later ages are not available for these children. In Chapter 3 we indicated that there was some evidence that rate of visual habituation at age 1 related to Stanford-Binet intelligence at $3\frac{1}{2}$. These results therefore provide, at best, indirect evidence for the effects of prenatal nutrition on subsequent intellectual development.

The results of the studies we have reviewed suggest that severe malnutrition in childhood occurring under the most adverse conditions and perhaps combined with the occurrence of chronic undernourishment might depress intelligence test scores. However, there is little available evidence at present that suggests that nutritional factors will account for any, or for any appreciable, variance in intelligence test scores in a representative sample of the U.S. population. And, there is little evidence to suggest that variations in intelligence test scores among different social groups are attributable to variations in nutritional status.

The Social Environment and Intelligence

To what extent are IQ scores influenced by the learning experiences provided for a person? Interest in this question goes back to Binet. It derives, in part, from the fact that scores in IQ tests are related to socioeconomic status. Children of affluent and educated parents tend to score higher on tests of ability than children of parents living in poverty who are not well educated. There are at least three kinds of explanations for the relationship between socioeconomic background and scores on tests. The relationship has been attributed to genetic differences in ability between different social classes, to differences in the adequacy of the biological environment in nutrition and health care which might relate to proper development of brain structures, and to differences in the cultural and learning experiences provided for children from different social classes. This section will deal with the influence of the nonbiological environment on scores on intelligence tests.

Evidence indicating that children from different socioeconomic backgrounds experience different learning environments and score differntly on tests does not indicate very much about the reasons for these differences. Children growing up in poverty may differ in genetic ability, may have been provided with inadequate nutrition, may experience a learning environment that does not adequately foster the development of their intellectual abilities, may develop attitudes toward tests and testers that lead them to perform at levels that do not reflect their true abilities, and may have knowledge and competencies that are not adequately sampled in tests because of a "middle class" bias in the content of tests.

Potentially, the least ambiguous source of information about the influence of a child's learning experiences on scores on intelligence tests comes from experimental studies in which there is an experimentally manipulated alteration in the environment. A simple clear-cut design involves a situation in which a group of individuals is randomly assigned to different experimental treatments. Usually one such group serves as a control group that does not experience any special environmental intervention. Other groups are provided some special environmental experience (typically involving extra stimulation) that is designed to increase cognitive ability. Differences in scores on tests between the experimental and the control group indicating that the former group(s) have made gains in intelligence test scores that are greater than those made by the control group permit one to infer that the experiences provided the experimental group do in fact increase scores on intelligence tests.

A number of such experimental studies have been conducted. For the most part these studies have dealt with the preschool child. Investigators have focused on the preschool period for a variety of reasons. They have assumed that intellectual plasticity is greater early in life than later. Therefore, programs that focus on the young child should have greater impact on subsequent intellectual development. Second, it has been assumed that the cognitive development of children growing up in poverty is impeded because of lack of cognitive stimulation. Therefore, early enrichment should permit the child to develop his cognitive potential and to profit from school experiences. Third, the initiation of Head Start programs with the development of day care centers in poverty areas provided researchers an opportunity to test the effects of cognitive enrichment on subsequent intellectual development.

Bronfenbrenner (1975) has reviewed a number of early enrichment studies based on preschool educational experiences. The typical study provides cognitive experiences in a preschool setting for a group of children coming from economically depressed families. The experiences extend for 1 or 2 years. And, a comparison of the experimental and the control group at the end of the experimental intervention (which typically ends at the start of kindergarten or Grade 1) may indicate that the difference in IQ approximated 13 points. These results suggest, at first blush, that it is not difficult to experimentally manipulate intelligence test scores—at least in young children. However, on follow-up there is typically a decline in the level of test score. And, by the time the children have reached the Grade 3 level (typically 2 or 3 years after the experimental intervention) the difference between the experimental and the control group has evaporated.

A typical study will illustrate this sequence of results. Weickart (1967) dealt with a group of black children living in Ypsilanti, Michigan, who had IQs between 50 and 85 on admission to the study and whose parents had little education and tended to be unemployed or in unskilled occupations. Children were randomly assigned to a control group or to an experimental group in which they were provided with a cognitively oriented preschool curriculum 5 days a week. At the end of the 2 years of intervention the experimental children had a mean intelligence test score of 94.7 and the control group had a mean score of 82.7. At the end of Grade 3, 4 years after the intervention, the mean test score of the experimental group was 89.6 and the mean test score of the control group was 88.1.

These typical results indicate that the changes that were attained do not persist after the intervention ends. It may be the case that continuing intervention would be necessary to maintain the initial gains. Also, there is some evidence that the initial gains may not have reflected a true or profound change in cognitive functioning.

The initial gains may reflect changes in a young child's capacity to feel at ease and to perform at his highest level in what may be, initially, a rather perplexing and perhaps frightening situation. Zigler,

Abelson, and Seitz (1973) have reported the results of two studies addressed to this issue. In the first study they administered the Peabody Picture Vocabulary test twice, 1 week apart, to a group of 4- and 5-year-old children living in conditions of poverty and to a group of affluent children. The IQ scores for the children living in poverty increased from 74.7 to 84.5 from the first to the second test administration. The children with middle class backgrounds increased their Peabody scores from 109.5 to 112.6. It is apparent that the mere act of retesting may lead to relatively large gains among young children with low scores on IQ tests who are unlikely to be familiar with the testing situation.

In a second study children from each of these backgrounds were assigned to one of four experimental conditions. In each condition children were given the Peabody test twice. The four conditions were obtained by studying situations in which the same examiner or a different examiner administered the test on the two different occasions with conditions in which the test was or was not preceded by a play situation. Table 5.7 presents the results of the experiment, which indicate again that the "disadvantaged" children made greater gains than the "nondisadvantaged" children. They also found the disadvantaged children had higher initial scores when their tests were preceded by a play period. No such effect was apparent in the nondisadvantaged children. These data indicate that some of the deficits in the early cognitive performance of children living in poverty are related to motivational influence rather than cognitive ability. A play period preceding a test or familiarity with the testing situation serves to decrease the apprehensiveness of the children and to increase their performance on tests by approximately 10 points. These findings help to place in perspective some of the gains attributable to early cognitive intervention.

Jacobsen *et al.* (1971) have reported a study that suggests that relatively large gains in test scores on the Stanford-Binet tests may be obtained after a relatively brief experimental intervention. They provided approximately 20 hours of training to 36 children. The children, who were mostly black, attended a day care center established for children whose families were below the federally defined poverty level. They were provided with training in solving two choice discrimination problems of increasing complexity. Some training in attention was included. The bulk of the training consisted of problem solving in which the correct solution was modeled (i.e., demonstrated by an adult), or reinforced (i.e., rewarded). There were 44 problems in all. Children in the experimental program were divided

TABLE 5.7

Mean Test–Retest PPVT IQ Scores (Study 2)[a]

Group	Disadvantaged				Nondisadvantaged			
	Test		Retest		Test		Retest	
	$\overline{X}$	*SD*	$\overline{X}$	*SD*	$\overline{X}$	*SD*	$\overline{X}$	*SD*
No delay, same *E*	81.83	12.43	88.58	8.46	113.83	4.87	116.25	5.08
No delay, different *E*	91.41	8.97	95.83	10.19	105.24	4.68	113.41	4.60
Delay, same *E*	78.50	7.23	88.99	6.30	110.33	9.50	115.58	4.08
No delay, different *E*	75.58	9.75	87.25	9.36	114.41	4.76	118.08	3.15

[a]From Zigler, E., Abelson, W.D., & Seitz, V. Motivational factors in the performance of economically disadvantaged children on the Peabody Picture Vocabulary Test. *Child Development*, *44*, Table 2, p. 299. Copyright © 1973 by The Society for Research in Child Development, Inc.

into three groups according to their initial IQ levels (the mean levels were 100.7, 89.2, and 72.8, respectively). At the end of the 20 hours of problem solving experiences the three groups had gained, respectively, 9.5, 9.7, and 20.1 points on the Stanford-Binet. The mean overall gain for the 36 children was 13.3 points. These data indicate that relatively brief but highly focused training can lead to relatively large gains in intelligence test scores in the preschool years. By the same token they suggest that test scores of children (particularly those who come from economically deprived families) may be depressed during the preschool years primarily because they lack fairly easily taught skills. Some of these skills probably involve simple test taking abilities—such as orientation to the tasks, paying attention to the task, and ability to understand what is required for successful problem solution. Thus low scores on intelligence tests in the preschool period may not reflect inadequate development of intellectual structures but lack of familiarity with test requirements. What is not clear is the extent to which a similar analysis may be applied to scores achieved in later childhood or in the adult years. Whimbey (1975) has published a book with the provocative title *Intelligence Can Be Taught.* In this book Whimbey advocates the use of "cognitive therapy" to increase scores on standardized tests of ability. He also briefly reviews a number of apparently successful attempts to increase intelligence test scores in adult samples. He cites four studies that provide evidence for the potential utility of cognitive therapy. One is an unpublished study by Marron indicating that prep school attendance tends to increase scores on the Scholastic Aptitude Test. Whimbey's brief description of the study does not make clear whether the study had an adequately chosen control group. A second study cited by Whimbey is a study by Bloom and Broder (1950) dealing with the improvement of academic performance among college students. Two other studies cited—one by Hardy (unpublished), the other by Whimbey himself—are really case histories indicating improvements in test scores for one individual in each case. The set of studies cited clearly does not provide convincing evidence that intelligence test performance can be reliably increased in the adult years by cognitive therapies. It may well be that intelligence test performance can be significantly increased in the adult years by considering test performance as the end result of a set of learned skills which can be taught in a reasonable period of time by special cognitively oriented therapies. An alternative position would hold that schools provide "cognitive therapy" and that individuals whose intelligence test performance is inadequate may be less tractable to cognitive therapy. In any case there is as yet no scientifically acceptable demonstration that the kinds of skills that are measured by intelligence tests can be taught to adults (or to children past the preschool years) in such a way that meaningful improvements in intelligence test performance can be obtained.

Rosenthal and Jacobsen (1968) have attempted to increase intelligence test performance experimentally by changing teachers' expectancies about pupil ability. They informed teachers of elementary school children in kindergarten and Grade 1 that some of their pupils had been identified by a test as "intellectual bloomers" and would show dramatic gains in intellectual ability during the school year. They reported that the pupils identified in this way had significantly larger gains in intelligence test score at the end of the school year than pupils in a control group. The Rosenthal and Jacobsen finding suggested that it was relatively easy to change intelligence test performance and that such changes could be attained merely by changing teachers' expectancies about performance. Presumably these changes in expectancy would lead teachers to treat children differently and this would lead to changes in pupil ability—perhaps through changes in pupil's self-image.

The Rosenthal and Jacobsen report has been severely criticized. One critic, Thorndike (1968), has indicated that the study would not have been published if it had been submitted to a scientific journal. The authors were able to publish their results in the form of a book which received extensive attention and citation in the popular literature. These are legitimate questions about the appropriateness of the statistical procedures used by Rosenthal and Jacobsen. In addition, the study has been replicated a number of times and the results have been quite consistently negative (e.g., Jose & Cody, 1971). It is therefore reasonably well established that merely changing teacher expectancies has little or no effect on performance on intelligence tests.

Our review of intervention studies leads to the pessimistic conclusion that we do not at present have techniques for changing intelligence test scores in a meaningful way by experimental intervention. The interventions we have considered are relatively limited in duration. Perhaps intelligence test scores can be changed only by extensive and dramatic intervention to alter the total cognitive environment of children and adults. An ambitious effort involving this kind of intervention has been undertaken by Heber and Garber (1971). They started with a group of 40 newborn children whose mothers had intelligence test scores less than 70. The children were randomly assigned to an experimental and a control group. The children in the experimental group were given a massive intervention program starting 2 weeks after birth. The intervention included maternal training and infant training by a team of psychologists who removed the child from the home in the morning and returned the child to its home late in the afternoon. The training started at 3 months of age. Prior to this the children were provided with "teachers" in the home several hours a day. In effect, Heber and Garber have provided for the total cognitive training of a group of children growing up in poverty conditions by removing the children from their home environment. The preliminary results are encouraging. The children in the experimental group have maintained (over frequent testing experiences) test scores of approximately 120. At the latest testing at 66 months, the experimental group had a mean IQ of 124. The control group had a mean IQ of 94. The 30-point difference reported is clearly of a larger magnitude than that obtained by preschool interventions of briefer duration (the usual increases being of approximately 12 or 13 points). In this sense the preliminary results achieved are encouraging. On the other hand a number of cautions are called for. First, the sample is small and the mean of 124 is obviously subject to sampling error. The sampling error of course can be in either direction. Second, and most important, results are not yet available for older ages. Most of the declines in intelligence test performance that resulted in other preschool intervention studies occurred during the first 2 or 3 years of elementary school. Thus the results obtained by Heber and Garber, while impressive, must still be regarded as preliminary until test results are available at later ages. It is possible that some regression of test scores will result in the experimental group. Alternatively, it is possible that the intellectual gains achieved are sufficiently profound that the children in the experimental group will be able to maintain or increase their level of performance.

While evidence for meaningful changes in intelligence test performances as a result of experimental intervention is at best ambiguous, there is reasonably clear evidence for large-scale changes in intelligence attributable to naturally occurring environmental changes. Perhaps the clearest evidence of such changes comes from studies of changes in intelligence with age. Cross sectional studies of the relation between age and intelligence have indicated a consistent decrease with age in test scores that has not been found in longitudinal research. These findings clearly indicate that intelligence test scores have been improving in the U.S. at a fairly rapid rate. Comparisons of the Army Alpha tests of World War I and World War II draftees as well as large scale testing since World War II suggest that the change in intelligence over the last 50 years in the U.S. can be conservatively estimated at one standard deviation. Precise estimates are difficult to arrive at because of the changing composition of test batteries and changes in the samples used in various studies. However, there is little doubt that there has been a fairly sizable increase in scores on standardized tests of intelligence. This increase is probably related to changes in educational level. What remains to be determined by subsequent research is the future trend of changes in intelligence test score performance. It is not clear if the relatively rapid increases that have occurred over the last 50 years will continue. Nevertheless the generational changes in intelligence that have occurred provide the clearest nonexperimental evidence for changes in intelligence test scores attributable to environmental influences.

6

Group Differences in Intelligence Test Scores

In this chapter we shall review studies that deal with the differences in intelligence test scores for groups of individuals who differ in group membership. In particular, we shall deal principally with studies of the racial differences in intelligence, and we shall comment briefly on studies of the relationship between family position and intelligence test score.

Race and Intelligence Test Scores

The use of intelligence tests as evidence for the presumed inferiority of various ethnic and racial groups goes back to Galton and has persisted throughout the history of the use of these tests. And, such views have contributed to the occasional unsavory reputation surrounding the tests. In recent years, of course, the interest in the study of intelligence tests has been given impetus by Jensen's (1969a) argument that the black–white difference in intelligence test scores is attributable, in part, to genetic differences in ability.

In what follows, we shall present Jensen's position, including a presentation of the evidence he cites in favor of his views, and we shall state our reasons for disagreeing with him.

Jensen (1973) has recently summarized and expanded his position on the genetic basis of differences in intelligence test scores between blacks and whites. We shall attempt to summarize the studies and arguments he advances for his position.

It should be noted that there is virtually complete agreement between Jensen and his critics in accepting the empirical finding that there exists a black–white difference in intelligence test scores. This difference is approximately one standard deviation—i.e., approximately 15 points in the conventional scoring system in which the mean intelligence quotient is 100 and the standard deviation is arbitrarily set at 15 (see Dreger & Miller, 1960, 1968; and Shuey, 1966).

Thus, the mean intelligence quotient for the black population of the U.S. on the usual tests standardized on white samples to have a mean of 100 is approximately 85. It should also be noted that the racial difference of approximately 15 is a difference in means. There is considerable variability around these means and a large number of blacks score higher than whites and score higher than the mean of the white sample. Approximately 15% of blacks score higher than the mean of the white sample. Accordingly, racial characteristics per se are not extremely powerful predictors of intelligence test score. However, the difference in scores becomes magnified in importance when one considers the differences in proportion of cases from the two populations having high or low scores. And, when extreme scores are related to socially relevant criteria, the importance of the average differences becomes greatly magnified.

For example, an intelligence test score of 70 is generally regarded as a cutoff that defines mental retardation—a level of intellectual functioning below which an individual requires special educational facilities and cannot profit from ordinary instruction in the classroom.[1] Such scores are approximately seven times more frequent per capita in the black population of the United States than in the white. Accordingly, if intelligence tests are used as a basis for a decision about classifying a child as mentally retarded, black children will be so classified in far greater numbers than white children. Similarly, blacks are far less likely than whites to have intelligence test scores above a high cutoff point. Thus, if intelligence tests, or tests highly correlated with them, are to be used as a basis for admission to professional schools, and if a high cutoff point for admission is set on

[1] We do not endorse the assertion that individuals with test scores below 70 require special educational facilities. We merely indicate that this is frequently asserted.

the tests, the probability of a black student's being above the cutoff point will be far lower than the same probability for a white person.

It should be noted that in at least one important respect Jensen has changed his position about the genetic basis for black-white differences in intelligence. In his 1969 paper, he introduced a distinction between what he called Level I and Level II ability. Level I ability is called associative ability and involves "the neural registration and consolidation of stimulus inputs and the formation of associations" (1969a, pp. 110–111). Level I ability involves relatively little transformation of input. It does not involve, in Spearman's terms, the eduction of relations and correlates. Level I ability is assumed to be distributed relatively homogeneously among different racial (and social class) groups. Level II ability or abstract ability is assumed to be measured by tests that are good measures of *g* and are "culture reduced." For example, Ravens progressive matrices and Cattell's culture-fair tests are presumably measures of Level II ability. Level II ability is assumed to be distributed unevenly among different racial (and social class) groups. Blacks and individuals of lower social class background are assumed by Jensen to be lower in Level II ability than middle-class white children. Jensen further assumed that the growth rates of Level I and Level II abilities are different among these groups.

Figure 6.1 presents the theoretical expectation of growth curves of these abilities for different kinds of individuals. Figure 6.1 indicates that the differences in Level II ability between middle- and lower-class individuals (and racial groups) will increase as a function of age, with the greatest difference occurring at asymptotic level at age 14. Thus, Jensen's original theoretical position was that there are genetically determined (or influenced) racial differences in the growth and development of abstract ability. And, furthermore, his position implied that on relatively pure measures of Level II ability (the Ravens Progressive Matrices Test) the magnitude of black–white differences in score would increase with age and time in school. Moreover, because schooling was assumed to have a relatively minor impact on a culture-fair test such as the Ravens, the expected increases in differences in scores on the test with age was assumed to be determined by the unfolding of a genotypically controlled phenotype that ought to be relatively independent of the quality and character of the educational experiences of different groups of individuals. Jensen (1969b) did in fact report some data that apparently supported this implication of his theory. Jensen (1969b) cited data collected by Rohwer reporting scores on the Ravens Progressive

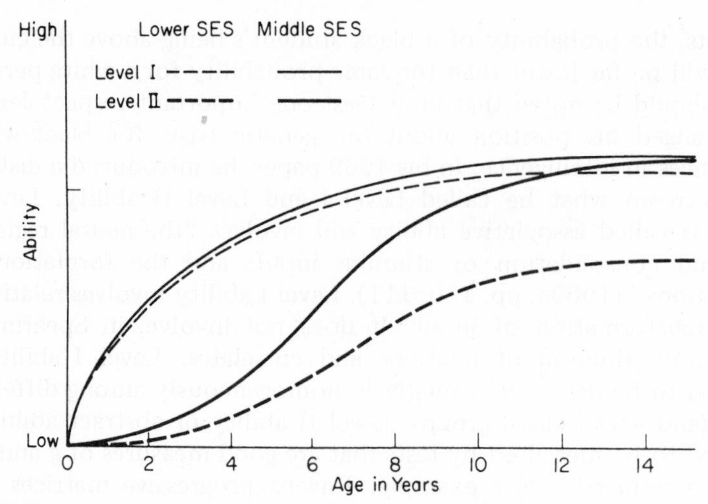

Figure 6.1. *Growth of Level I and Level II ability in two social classes. [From Jensen, A.R. Environment, heredity, and intelligence.* Harvard Educational Review, *1963,* 39, *116. Copyright © 1969 by President and Fellows of Harvard College.]*

Matrices Test for a group of children attending school in a lower-class black area and children attending school in an upper middle-class white area. He found that the black–white difference in Ravens scores increases from approximately 2/3 of a standard deviation at kindergarten to approximately 1 1/3 standard deviation units at the Grade 3 level.

Jensen (1971) subsequently presented data on the Ravens for a sample of children attending Berkeley public schools; they failed to support the notion of a progressively increasing black–white difference on the Ravens. Table 6.1 presents these data. It indicates that the black–white difference on the Ravens increased from .72 to 1.33 (in standard deviation units) from age 6 through age 8 but failed to indicate a progressive increase thereafter and subsequently declined from age 11 to age 12, dropping from 1.43 to .97 standard deviation units. Thus, the available data do not, in general, support the assertion that the difference between blacks and whites in tests purporting to measure Level II ability increases from age 6 to age 14. In any case it should be noted that Jensen has apparently abandoned this hypothesis. He now asserts (1974) that the differences between blacks and whites in standard deviation units remain invariant from the beginning of school through adulthood.

What data or reasons are advanced by Jensen in support of the

TABLE 6.1
Black–White Scores on the Progressive Matrices Test for Berkeley School Children[a]

Age range (in years) and months)		Number white	Number black	Mean difference (W-B/*SD* White)
5–5	6–6	91	64	.72
6–7	7–6	143	76	.77
7–7	8–6	79	66	1.33
8–7	9–6	96	42	1.35
9–7	10–6	83	55	1.59
10–7	11–6	71	45	1.43
11–7	12–6	69	27	.97
Weighted mean difference				1.12

[a]Based on Jensen (1971).

argument that the black–white differences in scores on intelligence tests are, in fact, attributable to genetic differences in ability?

Jensen's arguments are of three kinds. First, there are magnitude arguments. These are arguments that purport to demonstrate that the magnitude of known differences between blacks and whites in intelligence test scores is too large to be accounted for by known environmental factors. These arguments take for granted relatively large h^2 values and argue that the available environmental variance is not sufficient to account for differences of the order of magnitude obtained. Second, there are a series of arguments based on the presentation of various regression effects. Third, there are arguments based on the pattern of differences among various tests. These arguments purport to show that the largest black–white differences in test score are obtained for tests that are allegedly least influenced by culture and most heavily influenced by genotypes.

Magnitude Arguments

Jensen argues that the one standard deviation difference in intelligence scores between blacks and whites is too large to be entirely attributable to environmental influences in view of what is known about the heritability of intelligence test scores. He develops two specific estimates of the degree of environmental difference that would explain the existing racial differences in intelligence test scores. The first estimate is based on studies of separated MZ twins. Jensen relies on his analysis of the 122 known combined cases of separated MZ twins for which intelligence test data are available. The mean intrapair difference in test score for these twins obtained in four studies (see Chapter 5 for a review of these studies) is 6.60 points. Correcting for attenuation yields a true-score difference of 5.36 points. This estimate is compared to the estimated 15-point black–white difference in test score. The differences in twin pairs has a standard deviation of 4.74 points. This difference is entirely environmental. Also, the distribution of obtained differences is compatible with the assumption that the underlying distribution of environmental differences is normally distributed. This implies that an environmental difference between blacks and whites would have to be on the order of magnitude of 3.2 standard deviations to account for the difference in intelligence test scores.

Jensen goes on to note that the environmental difference is composed of both intrafamily and interfamily variance. Differences between MZ twins reared together provide an estimate of the intrafamily difference. Jensen assumes that approximately one-half of the environmental variance is variance within families. This implies that the interfamily standard deviation of environmental influences is 3.35 points. Therefore, the 15-point black–white difference in intelligence test score is equivalent to a 4.48 standard deviation difference in the environment.

A second method Jensen uses to derive an estimate of the magnitude of the environmental effect required to explain the black–white difference in intelligence is based on the assumption that h^2 = .80. The standard deviation of the distribution of environmental influences if h^2 = .80 is given by the following formula:

$$\sqrt{1-.80\,(.95)\,15^2} = 6.5,$$

where .80 represents the value for h^2, .95 is a correction for attenuation, and 15 represents the standard deviation (*SD*) of intelligence test scores.

If the standard deviation of environmental influences is 6.5, then 2.3 standard deviations are required to explain a 15-point difference in intelligence test scores. Assuming that half of the environmental influence is intrafamily variance, then the interfamily difference in environment required to account for the black–white difference would be 4.6. This value agrees well with the estimate obtained from the separated twin data.

Jensen argues that the known environmental differences between blacks and whites are not of the required order of magnitude, i.e., they are closer to one standard deviation than to four standard deviations. For example, he cites data (Jensen, 1973, p. 169) indicating that the mean difference in socioeconomic status between blacks and whites is .53 standard deviations in favor of whites. The difference in incomes is .80 *SD* units, .33 for unemployment rate, and .87 for children living with both parents.

In summary, the available data suggest that the environmental differences between blacks and whites are nowhere near the magnitude required to explain the differences between blacks and whites in intelligence test scores if the estimated interfamily environmental difference of 4.6 standard deviations is correct.

Jensen's magnitude arguments seem to us to be totally unconvincing and to use quantitative manipulation as a subterfuge for a balanced discussion of the issue. To begin with, the use of differences between MZ twins to derive estimates of environmental influences poses a number of difficulties. The available data (with the exception of Burt's data which, as noted earlier in our discussion of Kamin's critique of these data, is questionable) certainly suggest that the environmental differences between separated MZ twins do not represent a random sample of the possible differences between families. Separated MZ twins, at least where specific data are available, are almost invariably reared in families that do not differ greatly in social class background or cultural amenities. Therefore, the true differences between MZ twins reared apart in families that are unrelated or dissimilar in social class is undoubtedly greater than the differences used by Jensen to derive his estimates.

Also, we believe that the h^2 estimates used by Jensen are too high. And, as a result, the required interfamily differences in environment are smaller than those derived by Jensen.

Perhaps the weakest point of the magnitude argument is independent of the question about whether the estimated required magnitudes are correct or not. The argument contains the implicit assumption that black–white differences in environment can be **totally** represented by scales that are common to both racial groups. This analysis appears to be sociologically naive. U.S. society must be conceived of not only as a class system but also as having some of the properties of a **caste** system. To be black is to have environmental experiences that are not only quantitatively different but also qualitatively different from the experiences of whites. The type of data required to estimate the magnitude of the environmental influences on racial differences in intelligence can be derived from cross-fostering studies in which random samples of black and white children are raised in families representative of the opposite race. Since no such data exist—or, indeed, could be obtained under existing social conditions—any estimates of the influence of the environmental differences between races on intelligence test scores cannot be made. We would maintain that any conceivable evidence of the influence within races of environmental variations is an inadequate, indeed, virtually irrelevant, basis for estimating the magnitude of the environmental differences between the races. Indeed, for some of the very same reasons for which cross-fostering experiments are unthinkable, reliable estimates of the racial differences with respect to environmental variables that influence intelligence test scores cannot be obtained.

It should be noted that there exist data that, while only at best tangentially relevant, give some indication of the magnitude of influence of existing environmental variables. Perhaps the most relevant evidence of this sort comes from the Skodak and Skeels (1949) study of adoption. In their study they had intelligence test scores from 63 women who were the biological mothers of adopted children. The mean intelligence quotient for these women was 86. Although intelligence test scores were not available for the children's fathers, occupational data were available. As Kamin (1974, p. 131) notes, the occupational levels of the fathers were low—half were day laborers and one-quarter were semi- or slightly skilled laborers. Thus, a reasonable assertion about the expected value of the intelligence test scores of the biological father is 86. We can assume some regression toward the mean. The expected value of the biological child's intelligence if the child were reared by the biological parents would be between 86 and the population mean of 100. While precise values cannot be derived, a rough estimate of the regression toward the mean from the midparent IQ of 86 would be 93. This value assumes a correlation between midparent score and child's score of .5 and uses the standard regression formula, i.e., predicted score of child is equal to r times the z score for parents. It should be noted that the regression effect cannot be attributed solely to genetic influences. It represents the total parental influence on the child which is an amalgam of genetic and environmental influences. Thus, the expected value of the inherited genotype for intelligence must be less than 93. The extent to which it is less depends on the estimate of h^2. The children's intelligence quotient was 106—approximately 13 points higher than that which would be predicted on the basis of regression. If we assume, with Jensen, that $h^2 = .80$ and use his estimate of the interfamily standard deviation in environment of approximately 3.2 or 3.3 to account for the obtained difference between actual and expected intelligence test score, this would require in excess of 3.9 standard deviations between family environmental differences. The environments of the foster homes in which the children were reared, although undoubtedly superior to those that would have been represented by the biological parents, were not extraordinarily favorable. On a standard seven-point scale of occupational status going from professional Level I to day laborers (Level VII), the foster fathers averaged slightly higher than Level III (skilled trades) and the true fathers averaged between Level VI—slightly skilled—and Level VII.[2]

The Skodak and Skeels study suggests that a 15-point increment in intelligence test scores is a reasonable expectation of the expected influence of a middle-class environment on children with lower-class genotypes for intelligence. The magnitude of effect obtained is similar to the black–white difference in intelligence. Although it is a subjective matter, the magnitude of environmental differences between the biological and adopted families of Iowa children in the Skodak and Skeels study does not appear to us to be perceptibly larger than the black–white differences in environment.

Jensen presents additional data that he believes indicate that environmental characteristics are not sufficient to account for the magnitude of the racial difference in intelligence test score. Jensen (1973) notes that American Indians have been found to score lower than American blacks on virtually all indices of cultural and social background. The Coleman Report (Coleman, 1966), a report including a survey of the achievement of 645,000 American school children, found that American Indians scored lower than American blacks on all 12 environmental categories that were found to relate to academic success. Nevertheless, on all tests of achievement and ability from Grades 1 to 12, American Indians scored higher than

---

[2] Jensen, analyzing these data, assumed that the biological father's intelligence was 100, an unreasonable assumption in light of the occupational data. He then goes on to assert that the obtained difference is compatible with an environmental increase in standard deviation over the mean of 1.6 standard deviation units. In this analysis he used a standard deviation different from the value he uses in dealing with black–white differences. The appropriate *SD* to use is that for interfamily influences and on Jensen's own analysis should be 3.2 or 3.3. Thus, the observed differences in the Skodak and Skeels study between genetically expected and obtained values for children's intelligence tests are too large to be accounted for by Jensen's own model.

blacks. American Indians exceed American blacks on nonverbal and verbal tests and the differences are relatively large at least for nonverbal tests. At the Grade 1 level it is approximately on the order of one standard deviation. These data suggest that social deprivation, including such characteristics as inadequate health care, nutrition, and unemployment, are not sufficient explanation of the racial differences in intelligence. This follows from the finding that Indians, who are more socially disadvantaged in these characteristics than blacks, nevertheless score higher than they do on tests of achievement and ability. On the other hand, we do not believe that these data have much direct bearing on Jensen's genetic hypothesis to explain racial differences in intelligence. The principal limitation in these data is that they derive from a hyperselected group of American Indians. Approximately 75% of American Indians live on reservations and were not included in Coleman's study of school children. And, the American Indian families and children who have left the reservation may represent a sample of individuals who are biased with respect to intellectual ability.

Mercer (1973) has reported a study, based on a selection of a biased sample of American blacks, that has been interpreted as indicating that environmental variables can account for racial differences in intelligence. She studied five cultural variables related to WISC scores in a sample of Chicago school children. These were: living in a family with five or fewer members; having mothers who expected their children to be educated beyond the high school level; living in a family where the head of the family was married; living in a family that was buying or owned its home; and living in a family where the head of the household had an occupational index of 30 or above on the Duncan scale.

Mercer found that the mean IQ of her black sample ($N = 339$) was 90.5. For the sample of 17 black children with all five of the "modal" characteristics described above, the average IQ was 99.1. Mercer concludes that black children who come from family backgrounds comparable to those of the modal pattern of the national community have intelligence test scores comparable to the national norm. Of course this conclusion is based on a limited sample of children who are hyperselected. And socioeconomic deprivation is always ambiguous in causality. That is, social deprivation may cause low intelligence test scores and low intelligence test scores may cause social deprivation.

Loehlin, Lindzey, and Spuhler (1975) have indicated the ambiguity of such data in commenting on a reanalysis of the Coleman Report data presented by Mayeske *et al.* (1973). They assert:

> Writers in this area have often drawn rather strong conclusions based on a priori allocation of joint variance in one direction or the other. An example is provided by Mayeske *et al.* (1973, p. 126) in another analysis of the Coleman Report data. They show that most of the variance in student achievement that is predictable from racial-ethnic group membership could be predicted instead by a collection of other variables correlated with ethnic-group membership, including socioeconomic status, family structure, attitudes toward achievement, and the properties of the student body of the school the student attends. This amounts to saying there is a large joint component of variance that is causally ambiguous: the other variables could be predicting achievement because they predict racial-ethnic group membership; racial-ethnic group membership could be predicting achievement because it predicts the other variables; both could be predicting some third variable that in turn predicts achievement; or any combination of these in any degree could be involved. Mayeske and his associates, however, take only one of these possibilities into account in their interpretation and unhesitatingly allocate the joint variation to socioeconomic rather than racial-ethnic factors [Loehlin, Lindzey, & Spuhler, 1975, p. 166].[3]

It should be noted that the analysis reported by Mayeske *et al.* is somewhat more relevant than Mercer's analysis in that it includes data on a total sample rather than a hyperselected subsample. Nevertheless, the data implicitly refer to a subsample—that is, the subsample of black children who are least like the black sample and most like the white sample; and for this subset, differences disappear. The relevant issue raised by this sort of analysis is why are blacks who have the social backgrounds that produce high intelligence test scores a relatively small subset of the black sample? Perceptions will differ on this issue. We are inclined to believe that the appropriate answer lies in the history of social deprivation and racial discrimination which are part of the black experience.

Pattern of Black–White Difference in Scores

Jensen maintains that the magnitude of black–white differences in intelligence test scores differs on different tests. He asserts that the differences are largest in favor of whites on those tests that are the best measure of *g*, i.e., that come closest to measuring abstract intellectual ability. He asserts that the differences are smallest on tests that have the highest cultural loading. Jensen asserts that those tests that are the best measures of *g* are also the tests having high values of h^2. If it were true that tests that were more influenced by genotype and were the best measures of abstract intellectual ability

---

[3] From *Race Differences in Intelligence*, by John C. Loehlin, Gardner Lindzey, and J.N. Spuhler. W.H. Freeman and Company. Copyright © 1975.

showed the largest black–white difference, this would be evidence for a genetic hypothesis as a basis for the racial difference in intelligence test scores.

Jensen cites several studies in support of this assertion. The two studies he cites that provide relatively direct evidence for these assertions are a study conducted by him and a study by Nichols. Jensen obtained data for approximately 8000 children attending Berkeley, California, public schools (K through Grade 6) on a variety of tests including memory tests, intelligence tests, and a battery of achievement tests. He reports that the sibling correlations for these 14 tests are about the same among black and white samples. The sibling correlation ranges from .24 to .44 for the white sample and .15 to .45 for the black sample. Jensen uses these sibling correlations as a basis for estimating the heritability of the tests. The reasoning involved here is that the expected genetic correlation between siblings is approximately .5 (or slightly more if assortative mating exists). Deviations from .5 in either direction can therefore be taken as evidence of environmental influences. Jensen then relates the deviation from .5 for the sibling correlations for each of the tests to the difference between black and white means on each of the 14 tests. The correlation between the sibling deviation from .5 and the black–white mean differences is −.44 and −.34 for the white and the black sibling deviations, respectively. These correlations imply that the largest black–white differences are obtained for the tests for which the sibling correlation approaches .5. Nichols (1972) has reported comparable results for a different sample of 13 tests administered to black and white 7-year-olds. Using sibling correlations as an estimate of heritability he found that the correlation between estimated heritability of tests and the black–white difference in these tests was .67, indicating that the tendency of whites to do better than blacks was relatively strong for tests whose estimated heritability was high.

We have a number of criticisms of these data. First, the use of sibling correlations to estimate heritability is an almost ludicrously inadequate procedure. While it is true that low sibling correlations are incompatible with high values of h^2, sibling correlations of .5 (or higher, taking into consideration assortative mating) are compatible with either high or low heritability values. A sibling correlation of .5 could occur for a measure whose h^2 value is zero! These data would occur if there were influences present in the black environment that tended to depress scores on certain tests and to create intrafamily resemblance in scores for these tests. In short, these data are equally compatible with the assertion that the differences are due to genetic or environmental influences.

Quite apart from the illegitimacy of using sibling correlations as a basis for estimating heritability, Jensen's assertions that black–white differences are largest on tests that have high heritability fall down on a number of empirical and conceptual grounds.

There exist data that imply that differences between blacks and whites are larger or as large on tests that Jensen assumes are not highly heritable as they are on tests he does assume are highly heritable. For example, in a number of places in his 1973 book, Jensen suggests that the Ravens progressive matrices are one of the best measures of *g* and that this test is highly heritable. The Peabody Picture Vocabulary test—considered by Jensen to be a test that is not culture-reduced—is not a good measure of *g*, and is not highly heritable. Jensen (1971) reports data from his study of Berkeley school children that show the difference between white and black means for children ranging in age from 5 years, 5 months to 12 years, 6 months. Table 6.2 presents these data, clearly indicating that the differences between blacks and whites were not dramatically different on these two tests, and that they were slightly larger for the presumably less heritable Peabody than for the Ravens. Since these

TABLE 6.2
Black-White Differences in Two Tests of Intelligence at Different Ages in the Berkeley School System[a]

Age Range (in years and months)		Number whites	Number blacks	Test	
				Ravens W-B diff[b]	Peabody W-B diff[b]
5–5	6–6	91	64	.72	1.37
6–7	7–6	143	76	.77	.93
7–7	8–6	79	66	1.33	1.56
8–7	9–6	96	42	1.35	1.10
9–7	10–6	83	55	1.59	1.25
10–7	11–6	71	45	1.43	1.14
11–7	12–6	69	27	.97	1.58
			Mean:	*1.17*	*1.28*
			Weighted mean:	*1.12*	*1.24*

[a] Based on Jensen (1971).
[b] White mean–Black mean divided by white standard deviation.

data were available to Jensen when he wrote his 1973 book, and since the data came from his own research program, one wonders why he did not cite them in connection with his discussion of the relation between heritability of test scores and the magnitude of the black–white differences on the test.

These data are not anomalous. Jensen (1971) also reports data on the Lorge-Thorndike IQ test for a very large sample of black and white children in the Berkeley, California, schools. The differences on the nonverbal test at Grades 5 and 6 are 1.76 and 1.84 (in standard deviation units), respectively. The comparable differences on the verbal IQ for the same sample are 1.52 and 1.88, respectively. Presumably, the nonverbal part of the test is more culture-fair and a better measure of *g* (in this connection, see our discussion of the Crano, Campbell, and Kenny study in Chapter 4).

Also, Jensen (1971) discusses in the same article a study by Semler and Iscoe (1966) in which Ravens tests and the Wechsler Intelligence Scale for Children were given to a small sample of black and white school children in Texas. The black–white difference on the WISC at ages 7, 8, and 9 was .67, .52, and .59, respectively. For the same children the black–white difference on the Ravens Progressive Matrices test was .52, .36, and .20 for children at ages 7, 8, and 9, respectively. The black–white differences on the Ravens were slightly less than those obtained on the omnibus WISC which may be presumed to be a mixture of culture-reduced and non-culture-reduced items.

These data apparently indicate no support for the assertion that the black–white difference is largest on tests that are assumed, by Jensen, to be pure measures of *g* and to be most influenced by genotypes.

In our discussion of the question of magnitude of differences between blacks and whites on different types of tests, we have more or less explicitly accepted Jensen's framework of assumptions for the discussion. There are a number of assumptions made that are questionable. These include (*a*) the assumption that the Ravens test is a better measure of *g* than other tests; (*b*) the assumption that scores on tests such as the Ravens or the Cattell Culture-Fair test are more heritable than other tests of intelligence; (*c*) the assumption that scores on intelligence tests are equally heritable for blacks and whites; and (*d*) the assumption that the heritabilities of different tests are the same for blacks and whites.

Is the Ravens a better measure of *g* than other or different kinds of tests? Although this is a frequent assertion, in our judgment there is relatively little foundation for such an assertion. The assertion implies a kind of reification or hypostatization for *g* that belies its status as a statistical artifact. The composition of a *g* factor in tests is influenced in large measure by the tests that enter into the analysis. Thus, the precise loading of a test on *g* (i.e., the degree to which the test may be said to be a measure of *g*) depends on the composition of the factor analytic battery of tests. Further, our examination of Cattell's assiduous efforts to separate fluid from crystallized ability left us with the uneasy feeling that the separation was less than optimal and that at the more abstract levels of analysis the allegedly separate factors blended together again (see Chapter 2).

If the Ravens is not best conceived of as a measure of *g*, how may it be conceived? The British factor analytic school tends to conceive of the Ravens as a measure of the *k:m* factor, a factor defined by spatial and mechanical skills rather than verbal educational skills. These factors are extracted after the influence of *g* has been removed. Marolla (1973) has studied the relationship between the Ravens and socioeconomic and educational achievement in a cohort of 366,245 19-year-old Dutch boys taking their military fitness examinations. He finds that the Ravens is somewhat less related to educational achievement and socioeconomic background than tests of language skills used by the Dutch. However, what is of most interest in Marolla's research is that the Ravens is associated with the academic track selected. Individuals who score high on the Ravens were somewhat more likely to have entered the science and math track than the humanities track. Since the data were collected contemporaneously they do not indicate the causal direction of the relationship. That is, training in math or the sciences might increase the ability to do well in the Ravens, or high scores on the Ravens and the presumed spatial and visualization abilities measured by the test might predispose an individual to enter the science track. In either case, the Ravens may be conceived of not as a measure of *g* per se but rather as a measure of nonverbal abilities that are associated with interest, and perhaps capacity, to do well in the sciences and mathematics.

Other research has indicated that blacks tend to score somewhat lower in tests involving spatial and visualization abilities than in verbal abilities. Lesser, Fifer, and Clark (1965) have reported the results of a study comparing middle-class and lower-class individuals of different ethnic and racial groups in terms of their respective patterns of scores on the Thurstone Primary Mental Abilities Test. Figure 6.2 presents these results. Note that for each of the four

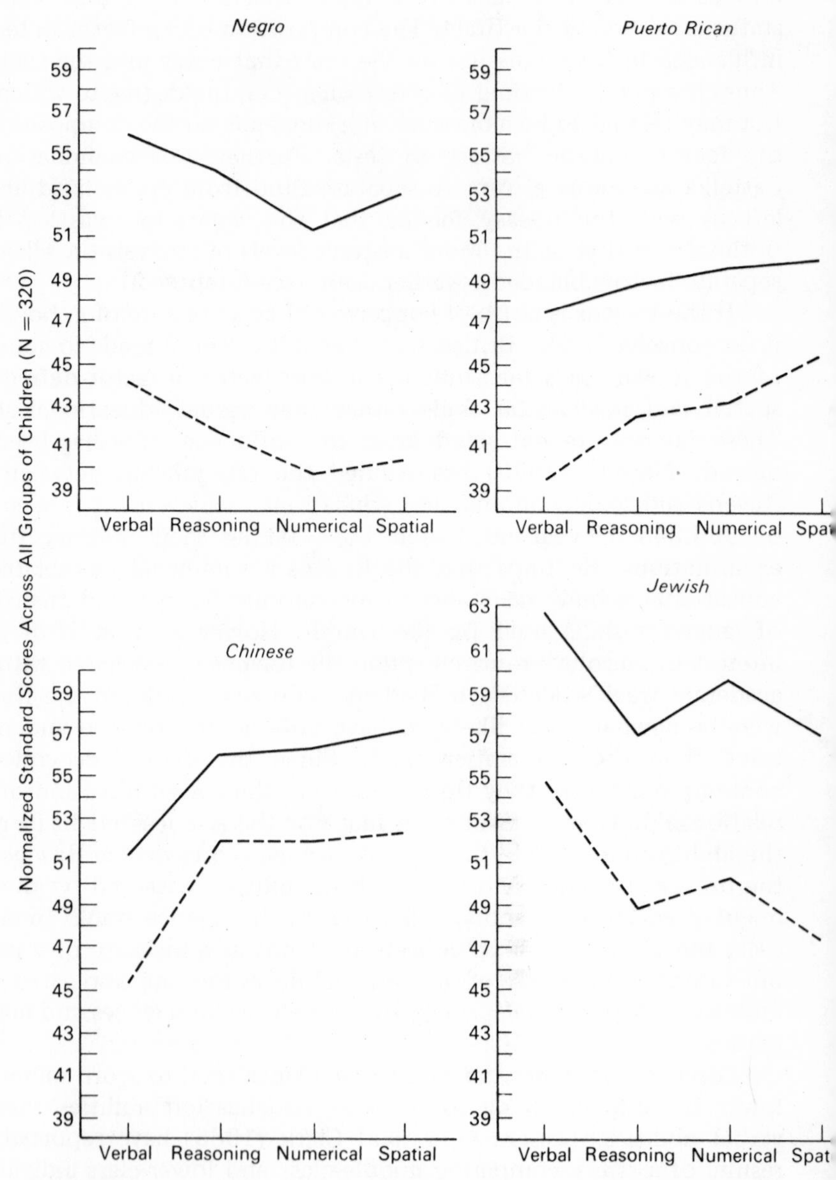

Figure 6.2. *Differing racio-cultural ability patterns. [From Lesser, F and Clark (1965).]*

ethnic groups high- and low-status individuals share the same pattern of scores across tests. The black scores are characterized by relatively high verbal scores and relatively low spatial and numerical scores. Note also that the pattern of high verbal and low spatial score is also characteristic of the Jewish group. This finding is of some interest in that Jensen (1973) asserts that Jews in America are an intellectual elite. He reports the results of a study indicating the per capita probability of individuals of different ethnic groups entering occupations that presumably require high intelligence (Jensen, 1973, pp. 252–253). Jews are far more likely, per capita, to enter such occupations than blacks or virtually any other ethnic group. Yet their pattern of scores (not their mean scores) is quite similar to blacks. This would appear to indicate that patterns of scores are not extremely predictive of intellectual competence although they may be associated with particular fields of interest. This, in turn, would imply that Jensen is unconvincing in his attempt to derive arguments for a genetic basis for black–white differences in intelligence from patterns of scores on different tests.

Let us assume that blacks tend to score low on tests of spatial ability and nonverbal tests (although the data do not invariably support this assertion).[3] Does this imply that blacks score lowest on tests for which scores are most heritable? The fact of the matter is that there are virtually no data indicating that scores on the Ravens or other comparable tests are more heritable than scores on verbal tests or more omnibus tests. Extremely high heritabilities have been reported by Burt for what is alleged to be Binet-type tests. Relatively little comparative data exist on this score. Cattell (1971) has reported h^2 values obtained using his Multiple Analysis of Variance Model for fluid and crystalized intelligence (see also Cattell, Blewett, & Beloff, 1955; Cattell, Stice, & Kristy, 1957). His preliminary estimates, which for a number of reasons he believes underestimate the value of h^2 for fluid ability, is that h^2 is .77 for fluid ability and .73 for crystallized ability—hardly a dramatic difference. In short, although many of hereditarian persuasion frequently argue that culture-reduced tests are more highly heritable than other tests of intelligence, the evidence for this position appears to be quite lacking. In summary, in our judgment the available data indicate that the differences between blacks and whites are of comparable magnitude on a variety of different kinds of tests. While there may be somewhat

---

[3] For a more extended discussion of this issue see Leohlin, Lindzey, and Spuhler, 1975.

distinctive patterns of abilities among blacks and various other ethnic groups, there is no evidence that differences in test score between blacks and whites are largest on tests for which h^2 is at a maximum.

Quite apart from the question of differential heritability for different kinds of tests, Jensen's analyses assume that intelligence test scores are equally heritable among blacks and whites. There is a lack of adequate data on the heritability of intelligence test scores for blacks. There are, however, four studies that deal with this issue.

Vandenberg (1970) studied black–white differences in heritability of a battery of 20 different cognitive tests in a sample of black and white twins. His black sample included 31 MZ and 14 DZ twins and his white sample included 130 MZ twins and 70 DZ twins. Within each racial group, he separately derived a test for the heritability of each of his 20 tests. The test involved the computation of an F ratio comparing the variance of differences between pairs of DZ twins to the variance of differences of MZ twins. The mean of the 20 F ratios of variance of DZ differences divided by the variance of MZ differences was 1.33 for the black sample. This overall mean was not significantly different from zero. This implies, for this particular sample, that DZ twins do not differ among themselves more than MZ twins. This, in turn, implies that there is no evidence of heritability for these tests taken as a whole in the black sample. The comparable mean F ratio for the white sample was slightly larger, 1.68, and this difference was statistically significant, indicating more evidence for heritability of these tests in the white sample.

The two mean F values for the white and black samples were not significantly different nor were any of the separate F values for each of the 20 tests significantly different.

It is not clear what is the appropriate conclusion to be drawn from Vandenberg's study. On the one hand the data do suggest, albeit weakly, that the heritability of ability tests may be lower among blacks than whites. Furthermore, the study is compatible with the assertion that intelligence tests may not be heritable at all among blacks. On the other hand, the study also fails to indicate that there is a significant difference in the heritability of ability tests among whites and blacks. In our judgment the study suffers from two limitations. First, the use of some combined or single index score would have been preferable and would have provided a more powerful index of intellectual ability for each child in the study. The relatively weak evidence for heritability in this study might, in part, be attributable to the use of 20 different scores with somewhat unstable characteristics. Second, and perhaps more severe, the black sample is relatively small, perhaps too small to provide a critical test of the heritability of intelligence in this group.

Osborne and Gregor (1968) reported a study of the heritability of spatial tests in a white sample composed of 140 MZ twin pairs and 101 DZ twin pairs and a black sample of 32 MZ and 11 DZ pairs. Values for h^2 were computed based on the comparison of intraclass correlations among MZ and DZ twin pairs. For nine separate tests, the h^2 values for the white sample ranged from .38 to .82 with a mean h^2 value of .54. In the black sample the values ranged from .02 to 1.76 with a mean value of .94. The null hypothesis of zero heritability could not be statistically rejected in seven of nine cases among the black sample (it was rejected in each case for the white sample). And, the average h^2 values were not significantly different between blacks and whites. The large nonsensical range of h^2 values in the black sample and the failure to detect a statistically significantly difference between racial groups in heritability undoubtedly stems from the inadequate size of the black sample.

Scarr-Salapatek (1971) has reported a study with adequate sample size for dealing with this issue. Her study involves a comparison of black and white twin data obtained by an examination of records of children attending public schools in Philadelphia. From this sample she obtained all children identified as twins. Since the sample of same-sex twins contains both MZ and DZ twins (the proportion of MZ twins can be roughly estimated) and the sample of opposite-sex twins is composed exclusively of DZ twins, we would expect same-sex twins to be more similar to each other in test score than the opposite-sex twins if the scores are influenced by genotype. This follows from the expectation that the MZ twin pairs will have smaller differences due to their shared genotype and thus the average difference between same-sex twin pairs will be smaller than the average difference between opposite-sex twin pairs. Thus, these expected differences provide a **crude** test for heritability. The test is crude because the exact proportions of MZ and DZ pairs among the same-sex twin pairs can only be estimated. Also, if there are environmental factors that tend to make opposite-sex DZ twin pairs less alike than same-sex DZ twin pairs, then the test is not valid. However, the available data do indicate that opposite-sex DZ twins are about as similar in tests of ability as same-sex DZ pairs (see Chapter 4).

Having obtained a sample of same- and opposite-sex twin pairs, Scarr-Salapatek then obtained information on group test performance. Different tests were used at different grade levels and some statistical adjustments were required to make the test scores comparable.

These data were used to compute intraclass correlations for the two types of twins on a combined measure of ability. For the total black sample the intraclass correlation for 334 same-sex twins was .57 and for 169 opposite-sex twin pairs it was .59. Thus, among the black sample there was no evidence whatsoever of a genetic influence on ability. The same analysis for the white sample produced intraclass correlations of .753 for 193 same-sex twin pairs and .694 for opposite-sex twin pairs, indicating some weak evidence for heritability of total ability in this sample.[4]

In addition, to an analysis by race alone, the data were analyzed by both race and class. Table 6.3 presents these results. What is striking in the data in Table 6.3 is the virtually complete absence of evidence for heritability of scores among individuals below median social status level in both black and white groups. For the advantaged groups there was some evidence of heritability in that the correlations for same-sex twins were somewhat higher than the correlations for opposite-sex twins. The Scarr-Salapatek data taken as a whole imply that race is not the critical determinant of heritability of tests but that social status is. Since there is a social class difference between blacks and whites (in favor of whites), it would imply that intelligence tests are less likely to reflect genotypic capability among blacks than among whites.

The Scarr-Salapatek study raises the fundamental issue of the difference in heritability of tests among different groups. Although the study is suggestive, it cannot be considered definitive. It is certainly adequate in sample size. The principal fault with the study is the failure to use direct measurement of zygosity. Thus, heritability is at best only weakly inferrable from such a design.

Nichols (1970) has reported a twin study involving a comparison of blacks and whites on the heritability of intelligence in a group of

[4] The results for the verbal and nonverbal components of the total score were somewhat different. Black same-sex twins had correlations on the verbal and nonverbal components of the combined index of .54 and .54, respectively, and black opposite-sex twins had correlations of .44 and .49 on these measures. Among whites the same-sex twins had correlations of .70 and .63 on the verbal and nonverbal tests, respectively, and opposite-sex twins had correlations of .59 and .66 on these data. Thus the data for the separate component scores indicate somewhat more comparability in heritability between racial groups than the total score. We have emphasized the total score in the belief, possibly erroneous, that it is a more valid index. Scarr-Salapatek believes that the verbal tests were more comparable across grades and should be considered the best index.

TABLE 6.3
Intraclass Correlations among Black and White Twins in Different Social Groups on Different Tests[a]

	Black		White	
	Same sex	Opposite sex	Same sex	Opposite sex
		Verbal		
Below median	.49	.42	.49	.55
N = (211)		(117)	(41)	(16)
Middle or above	.62	.46	.68	.55
N = (123)		(62)	(153)	(70)
		Nonverbal		
Below median	.51	.52	.52	.62
Middle or above	.57	.44	.63	.63
		Total		
Below median	.53	.60	.60	.63
Middle or above	.63	.57	.75	.65

[a]From Scarr-Salapatek, S. Race, social class, and I.Q. *Science, 174,* 1285–1295, Table 4. 24 December 1971. Copyright 1971 by The American Association for the Advancement of Science.

4-year-olds as measured by the Stanford-Binet. He found correlations of .62 and .51 for 36 and 55 pairs of white identical and fraternal twins, respectively. The comparable data for black identical and fraternal twins were .77 and .52 for 60 and 84 twin pairs, respectively. These data suggest higher heritability for blacks than for whites. However, the data must be accepted cautiously for several reasons. Intelligence tests at age 4 are not as reliable as those given later in life. The correlation for white MZ twins (.62) is extremely low and is deviant from the comparable values reported in the literature. For example, Wilson (1975) has reported a correlation of .82 for white MZ twins on the WISC given at age 4. Nichols indicates that data on sibling resemblance on the 4-year-old Binet indicate lower similarity among black siblings that white siblings (r = .37 for 970 black sibling pairs, and r = .52 for 1100 white sibling pairs). Nichols argues that these data imply, or are compatible with the view, that there is lower heritability for Binet scores in the black sample.

The available data on the heritability of intelligence in black samples is nowhere near as extensive as the data for white samples. There are some data, principally the Scarr-Salapatek study, that suggest that heritability may be lower in black samples than in white samples, but other data do not support this finding. In any case, good data on the heritability of intelligence among blacks is a precondition for dealing with black–white differences in intelligence in genetic terms. That is, high heritability of intelligence test score among blacks is a necessary, but not sufficient, condition for a genetic explanation of black–white differences in intelligence. In the absence of such data, a consideration of a genetic basis for black–white differences in intelligence appears premature.

Differences in heritability for tests among different social groups are relevant to what Jensen (1969; 1973) has called the threshold hypothesis. The hypothesis assumes that below some level of environmental adequacy, genetic potentials for intellectual development tend not to be realized. This implies that heritability for test scores will be lower in "deprived" groups and that changes in the environment will have larger effect for "deprived" children than for children above the threshold. Jensen speculates that going from a lower class to an average environment might lead to greater change than going from an average to a superior environment. Assume that the threshold hypothesis is correct. Assume further, as is generally conceded, that blacks are as a group more likely than whites to be below the threshold of adequacy. These assumptions imply that some or all of the black–white differences in intelligence test scores may be attributable to the possibility that intelligence tests do not reflect the true genetic potential of many blacks. No one should doubt that some version of the threshold hypothesis is correct. Clearly, there are some environments that are so destructive of the development of intellectual capacity that scores on tests are useless for inferring anything about a possible genotype for intelligence. What is at issue is not the correctness of a threshold hypothesis but a quantitative question of the appropriate location of the threshold. A "high" threshold model suggests that the appropriate level of environment required for the fostering of genetic potential is one that is not likely to be met by a relatively large proportion of individuals and in particular is a level not likely to be experienced by many blacks. A "low" threshold theory would suggest that most individuals in our society experience a sufficiently adequate environment for the development of their ability and that measured test scores, to a relatively equal extent, reflect underlying ability for most individuals. In particular, most blacks are above the necessary level. We have very little data that would permit us to obtain a better insight into an appropriate version of a threshold theory. Such data are badly needed. The Scarr-

Salapatek study, despite its shortcomings, provides virtually the only relevant data directly dealing with this issue. And, the study suggests that something akin to a high threshold theory may be correct. This, in turn, would suggest that black–white differences in intelligence test scores tend to be due to the probability of experiencing environments that are adequate to the fostering of intellectual ability.

Regression Arguments

Jensen's regression arguments are of two types, regressions among tests and regressions among different groups of individuals. Jensen's regression arguments among individuals attempt to derive predictions from the assumption that there is a true genetic difference in intellectual ability between blacks and whites. If this were so, we would expect blacks who are high in intelligence test scores to have siblings or children who regress farther toward the lower genetic mean. That is, if the true genetic mean for blacks is 85 on intelligence test score, then the sibling of a black person with a high intelligence test score, say 120, can be predicted to have a test score between the value of his sibling and the black mean of 85. A white person with the same IQ can be expected to have a sibling whose test score is between 120 and the presumed white mean genotype of 100. Thus, the black individual's sibling would show larger regression toward the mean. Jensen reports the results of such a regression study using his Berkeley data. Jensen (1971, p. 118) reports that if black and white children are matched for intelligence test score, the full siblings of the black children will average 7 to 10 points lower than the siblings of matched white children. For example, he asserts that black children with intelligence test scores of 120 have siblings whose average intelligence test score is 100. Correspondingly, black children with intelligence test scores of 70 will have siblings whose average score is about 78. White children with a score of 70 will have siblings with a score of 85.

Differential regression effects have also been noted for black and white children of different social class. Shuey (1966) noted that black–white differences in test score were somewhat larger for upper status groups than for low status groups. That is, if blacks and whites are matched for social status, the differences in test score are larger between high status than low status groups. Scarr-Salapatek (1971) has reported similar findings. Among low status groups in her study the black–white differences in total ability score expressed in standard deviation units (Mean White-Mean Black/Standard Deviation of Whites) was .42 and for high status groups the comparable difference was .89. Jensen explains these findings by appeal to the notion that blacks are regressing toward a lower genetic mean than whites.

In addition to tests of regressions between individuals, Jensen (1973, pp. 306–312) also discusses a study dealing with differential regressions between test scores. Jensen examines predictions about regressions derived from different hypotheses about differences between blacks and whites in test scores. Suppose the differences were due entirely to environmental influences. If this were the case, a particular phenotypic score among blacks would imply a genotype higher than the same phenotypic score among whites because the phenotypes had been artificially depressed. Similarly, the same genotypes for blacks and whites would be associated with different phenotypes. Blacks with the same genotypes as whites would tend to have lower phenotypic scores. Assume furthermore that tests of intelligence differ in the extent to which they are measures of an assumed underlying genotype for ability. This would permit the comparison of regressions of a test that comes closer to measuring a genotype for ability (the Ravens) with a test that is assumed to be less reflective of genotype (the Peabody), and vice versa. The expected regression effects implied by the environmental assumptions are the following: Equally high scores on the Peabody test (the more purely phenotypic measure less influenced by genotype) would always be associated with higher scores on the Ravens (the allegedly more purely genotypic measure) for blacks than whites. This follows from the assumption that environmental disadvantages will have a larger impact on the Peabody than on the Ravens. Hence the same phenotypic score on the Peabody for blacks and whites would imply higher genotypic scores on the Ravens for blacks. Conversely, the predicted effects of the regression of Ravens scores on Peabody scores are different. For equal Ravens scores, blacks should always have lower Peabody scores under these assumptions.

Figure 6.3 presents regression analyses performed by Jensen indicating the regressions of Ravens and Peabody scores for blacks, whites, and Mexicans. The differences in regression between whites and Mexicans reported in Figure 6.3 support the environmental explanation of the differences in test scores between these two groups. That is, for Mexicans the same scores on the Peabody are associated with higher scores on the Ravens than for whites, suggesting that the phenotypic score is depressed by environmental influences. Conversely, equal scores on the Ravens for Mexicans are

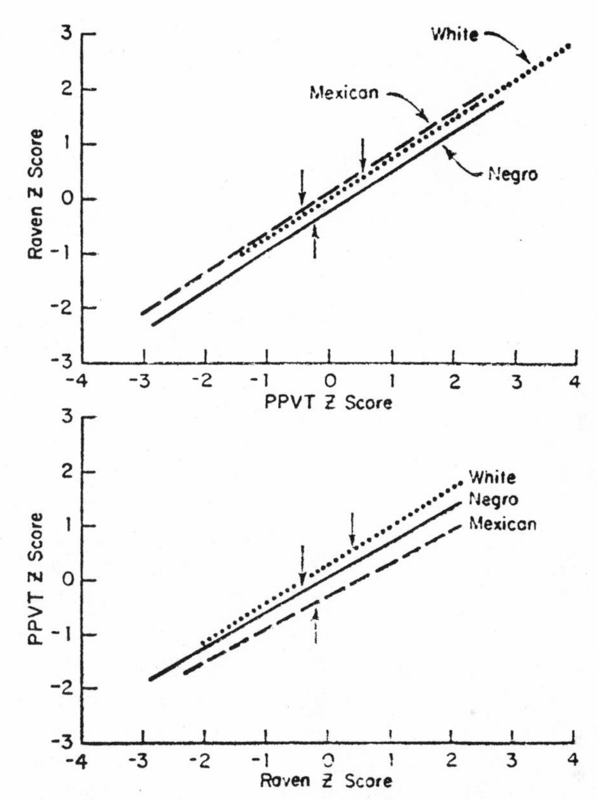

Figure 6.3. *Regression of Raven's Matrices standardized scores (z) on Peabody Pictur Vocabulary Test z scores (top) and regression of PPVT scores on Raven scores (bottom). The vertical arrows indicate the bivariate mean of each group. [From Jensen (1973).]*

associated with lower scores on the Peabody than for whites. However, the differences in regression between these two tests do not support an exclusively environmental hypothesis of the difference between blacks and whites on these two measures. The comparison of black–white regressions on these measures are in fact compatible with a hypothesis that implies that blacks are lower than whites in both genotype and environment and, in addition, that the magnitude of genotypic differences is larger than the magnitude of environmental differences.

We do not find Jensen's regression arguments particularly convincing. Furby (1973) has indicated that it is fallacious to interpret differential regression effects as evidence for a genetic hypothesis. In effect, Jensen's arguments about regression are valid only if one assumes that his theory is correct in the first place. That is, Jensen is indeed correct in his assertion that siblings of blacks with high test scores will have lower test scores than siblings of whites with equally high test scores if the true average genotype for blacks is lower than that for whites. However, the same deduction can be derived from a purely environmental hypothesis. If the differences between blacks and whites in intelligence test scores are due entirely to environmental factors, then we would expect that siblings of black children who score high in intelligence tests would have lower scores than siblings of white children who score equally high in intelligence tests. Regression toward the mean is statistically necessary independent of the reasons for the differences in the mean. We can illustrate this notion. Assume that intelligence test scores (for both blacks and whites) are totally determined by environmental influences—i.e., h^2 = zero. Assume that the environmental influences on intelligence test score to which blacks are exposed are such that the average score is 85. Assume further that a particular black child scores 115 on an intelligence test. This child has experienced an environment that is unusually favorable in its influence on intelligence test scores among blacks. (Assuming a standard deviation of 15, his z score value for his environment would be $+2z$.) We can assume that the sibling of such a black child would also be likely to experience an environment that is more favorable for the development of intelligence than is typical for other black children. However, as long as the environments of siblings are not perfectly correlated, we would assume on the average that his environment would be less favorable than that of his sibling. If we assume an intersibling correlation in environment of .5, then the expected or average value of siblings of black children who score 115 on intelligence test scores would be derived from the standard regression formula as follows:

$$\text{Predicted } z \text{ score of sibling}_2 = r \cdot z \text{ score of sibling}_1$$

If sibling 1 has a z score of 2, then the predicted z score for his sibling would be 1, implying an average IQ of 100 for all black siblings whose siblings have intelligence test scores of 115. The comparable predicted white regression score would be 107.5. Thus, a purely environmental analysis also implies differential regressions. The underlying empirical state of affairs associated with the environmental differential regression concept is simply that the blacks are exposed to an environment that is, on the average, less favorable in its influence on test scores than is the environment to which whites are exposed. If a particular black child is exposed to an environment that is unusually favorable for the development of intelligence, this is a statistical rarity. His sibling, while on the average experiences an above-average environment for a black, is less likely to experience an equally advantageous environment, and is more likely than his favored sibling to be exposed to some disadvantageous environmental influences which are more characteristic of the black experience than of the white experience. Hence his score must, of statistical necessity, regress toward the black mean. The same analysis also relates to parent–child regressions. Black parents with high intelligence test scores can be predicted to have children with lower test scores than white parents of equally high intelligence test scores. This is analogous to the assertion that an atypical black is a black anyway. The factors that influence his children's test scores (or any other characteristic, for that matter) will be an amalgam of special factors conferred upon the child by virtue of the special characteristics of the parents, and influences on the child obtained from his group membership, which are independent of the parental influence. In summary, differential regression toward different group means must occur independent of the reasons for the group means in the first place. Jensen's arguments about differential regressions support his genetic theory about group differences only if one assumes that the theory is correct in the first place. And, differential regressions among siblings or between parents and children are simply irrelevant as evidence for a genetic hypothesis about group differences.

In addition, we find Jensen's arguments about differences in regression among tests equally unconvincing. Jensen's study of regressions for different groups on the Ravens and the Peabody bears on the issue of a genetic hypothesis of group differences only if one assumes that the environmental influences, if any, that depress scores for blacks on intelligence tests influence the Peabody test more than the Ravens. We have already discussed this issue in connection with our discussion of the Ravens and we have rejected this notion. The analysis of the Mexican–white differences is more persuasive. That is, since Mexicans are bilingual, one would expect them to do poorly on a verbal test such as the Peabody, particularly a test involving knowledge of words that are statistically rare in English.

In addition to differences in the mean intelligence test scores of blacks and whites, there is evidence that blacks and whites differ in intelligence test scores in at least three other respects. These are: variance differences, sex differences, and differences in the relation between ability and achievement.

Shuey (1966) has summarized studies that involve comparisons of variance differences in intelligence test scores between blacks and whites. Among 200 studies permitting such a comparison on the same tests, 67% of the studies reported significantly larger variances for whites than for blacks; 26% showed the opposite. Kennedy, Van De Reit, and White (1963) in their large normative study of the Stanford-Binet in the southeastern United States found that the black variance was 57% the size of the white variance. Although the data are not definitive, there is some indication of lower variance among scores in the black sample. The somewhat lower variance in intelligence test scores that is occasionally found among black samples may be attributable to any of a number of different reasons. Since blacks are far more likely per capita than whites to be placed in special classes for the retarded or to be placed in institutions for the retarded, it is possible that some of the studies may suffer from sampling bias. That is, a larger percentage of blacks than whites with very low scores were excluded from the sample, thus reducing variances. It is logically possible that there is less genotypic variance among blacks. For example, if assortative mating for intelligence was less among blacks (and/or intelligence test scores were less influenced by genotype among blacks), then there would be less genotypic variance among blacks. There may be less environmental variance for factors that influence test scores among blacks than among whites. Or, the covariance between genotype and environment may be less among blacks. If many blacks experienced environments that were not conducive to the development of the ability to score high on tests, then test scores among blacks would be less likely to reflect their genetic potential. Accordingly, there would be a decrease in the tendency for favorable genotypes to be associated with favorable environments. This would serve to reduce covariance and lead to a decrease in the overall variance of black scores. At the present time there is little or no reason to choose among these possible explanations for the possible reduction in variance in test scores.

Jensen (1971) has summarized the available data on the magnitude of sex differences between blacks and whites on a variety of tests of intelligence. Table 6.4 presents the results of his analysis, which indicate that there is a slight tendency for black males to do somewhat more poorly on tests of intelligence than black females. The sex difference among blacks is somewhat larger than the sex

TABLE 6.4
Sex Differences among Blacks and Whites on Tests of Intelligence[a]

Study	Test	N		M–F[b]	
		Black	White	Black	White
Jensen (1971)	Peabody	374	631	.29	.42
Jensen (1971)	Ravens	375	632	.11	.18
Baughman and Dahlstrom (1968)	Binet Primary Abilities	1184	920	−.21	.08
Wilson (1967)	Henmon-Nelson	4707	5817	−.11	−.07
Semler and Iscoe (1966)	WISC	134	141	.27	−.07
Semler and Iscoe (1966)	Ravens	82	78	.08	.14
Jensen (1971)	Lorge-Thorndike	4001	5256	−.19	−.08
Jensen (1971)	Figure copying	2250	2732	−.17	−.10
Jensen (1971)	Memory	2134	2615	.20	−.16
Weighted Mean				−.07	−.13

[a] Based on Jensen (1971).
[b] Male mean–female mean/*SD* white.

difference among whites. The exact reason for this social difference between the sexes is not known. The very small difference that exists in this respect tells us relatively little about the reasons for the relatively large racial differences in mean scores.

In Chapter 4 we indicated that it was not until recently that data were obtained that dealt with a fundamental assumption of what is measured by intelligence test data. Crano, Kenny, and Campbell (1972), using cross-lagged panel analysis, were able to show that intelligence tests did in fact measure something causally related to achievement and, in addition, they were able to reject the alternative causal hypothesis that achievement was causally related to subsequent intellectual ability. The analysis we reported dealt only with suburban school children. They also had a large sample of children who attended school in the city of Milwaukee. This sample was predominantly black. For this latter group of children there was no evidence that intelligence tests were causally related to subsequent achievement. For example, they found that the correlation between intelligence test score at Grade 4 and a composite measure of achievement at Grade 6 was .61. The correlation between achievement at Grade 4 and intelligence test score at Grade 6 was .62. In addition there was no evidence in their sample that nonverbal Lorge-

Thorndike scores (presumably reflecting fluid ability) were causally related to verbal Lorge-Thorndike scores. These data indicate that the causal networks that are central to the validity of intelligence test scores and bear fundamentally on the adequacy of the interpretation of the score as a measure of the construct ability appear to break down and not to be applicable to a predominantly black sample. This in turn suggests that intelligence tests are not a measure of the same construct in white economically advantaged samples and in black economically disadvantaged samples.

Studies of racial hybrids provide another source of data that bears on the question of the reasons for black–white differences in intelligence. Shockley (1971a, 1971b, 1971c, 1972) has proposed a test of the genetic hypothesis to explain racial differences in intelligence. He proposes that correlations be obtained between the frequency of certain blood groupings among blacks and the intelligence test scores of blacks. Shockley argues that certain blood groups are more characteristic of whites than blacks. Therefore, the presence of such blood groupings within a black person indicates that the person is likely to have had a white ancestor and to be a "social hybrid." Such evidence is presumably objective and is far less contaminated by possible social reactions to more overt indices of mixed ancestry such as skin color. If the presence of blood groupings indicative of white ancestry among blacks is associated with higher intelligence test scores, then Shockley argues that evidence would exist in favor of a genetic explanation for racial differences in intelligence.

We believe that Shockley's research proposal is both biologically and sociologically naive. And, the one study known to us that is relevant to the hypothesis has failed to find an association between blood groupings and intelligence in a black sample. The research design is sociologically naive because it fails to consider the possible social influence of a white ancestor during the period of slavery when most racial admixture apparently occurred. Racial inheritance may often have been, at least historically, not unrelated to social advantage. Biologically, the presence of blood groupings indicative of white genes may become, with the passage of time, less correlated with the genes that are indicative of intelligence. In this connection, Loehlin, Vandenberg, and Osborne (1973) found that the presence of blood groups indicative of European ancestry among blacks was uncorrelated with the presence of other blood groups indicative of European ancestry. In two separate black samples they report average correlations between blood groups indicative of European ancestry of −.04 and −.02, respectively. These data indicate that genes indicative of European ancestry are now unrelated. Accordingly, it is reasonable to assume that such genes are not highly correlated with the presence of white genes related to intelligence.

Loehlin, Vandenberg, and Osborne (1973) have reported two nonsignificant negative correlations between the presence of genes indicative of European ancestry and intelligence test scores in two samples of black children. Loehlin, Lindzey, and Spuhler (1975) cite two studies of "racial hybrids" that do provide evidence in favor of an environmental interpretation of racial differences in intelligence. Eyferth (1961) obtained intelligence test data for illegitimate children in Germany who were racial hybrids with white mothers and black fathers. The majority of the fathers were U.S. servicemen. They compared the test scores of these children with those of illegitimate white children who were matched for age and mother's social background. There was no difference in test scores for these two groups of children. They hybrid children clearly had a significant number of black genes and, if genetic differences in racial ability existed, one would expect that the children of black fathers would have lower test scores. The Eyferth study is not absolutely conclusive because data on the actual background and test scores of the fathers were not available. However, it is known that there was approximately a one standard deviation difference in intelligence test score for black and white soldiers during World War II (Davenport, 1946). Such a difference should have resulted in a difference in the test score of their children on a genetic hypothesis. In order to argue that these data do not contraindicate a genetic hypothesis, one would have to assume that black and white soldiers who fathered illegitimate children in Germany after World War II were a nonrepresentative sample of black and white soldiers with respect to genetic intellectual ability. In particular, one would have to assume that the black soldiers who fathered such children were unusually high in genetic intellectual ability relative to other black soldiers and/or that the white soldiers who fathered illegitimate children were unusually low in genetic ability relative to other white soldiers. While this is possible it is not highly likely. Alternatively, these data are compatible with, and supportive of, an environmental interpretation of the racial difference in intelligence test score.

Loehlin, Lindzey, and Spuhler (1975) have presented another way of using data on racial hybrids to test hypotheses about the genetic and environmental basis for differences in intelligence test score. Their analysis focuses on black children with high test scores. As we have seen, relatively small differences in means in two distributions will lead to relatively large differences in the proportion of cases found at the extremes of a distribution. If there is an association between genes characteristic of racial membership and genes that influence the development of intelligence, this association ought to lead to large differences at the extremes of the distribution. Accordingly, one would expect in a genetic hypothesis explaining racial differences in intelligence that, among black children, there should be a disproportionately large number of racial hybrids who score above a high cutoff point in intelligence test score. The probability that a black child in school will have an intelligence test score $\geq$140 is .0012 (Shuey, 1966). Witty and Jenkins (1936) were able to find 28 black children with test scores in this group. They obtained genealogical data for these children and they found that the distribution of degree of black ancestry was about comparable to that obtained for the black population as a whole, using comparable methods of determining the degree. These data fail to support genetic explanations. If the genetic hypothesis were correct, one would expect that the small number of black school children with unusually high test scores should have a disproportionately large percentage of white ancestry.

There are other data on racial hybrids that have been cited in discussions of racial differences in intelligence which, in our judgment, are not particularly critical. DeLomos (1969) has reported a study dealing with Australian aborigines. He compared the performance of children who were completely aboriginal with that of children growing up in the same cultural setting who were predominantly "7/8" aboriginal, using Piagetian tests of conservation as a measure of intellectual ability. He found that the partly aboriginal children performed better than the children who were completely aboriginal.

For at least three reasons, DeLomo's findings are not critical. First, they deal with Australian aborigines and, irrespective of the validity of the findings, should not be generalized to the black population of the U.S. Second, there may be subtle environmental differences associated with partly European ancestry that are responsible for the obtained differences. Third, using some of the same subjects and similar tests, Dasen (1972) has failed to replicate DeLomo's findings.

Willerman, Naylor, and Myrianthopoulos (1974) have tested the intelligence of 4-year-old children with mixed white and black parents. They report that children with white mothers and black fathers score approximately nine points higher on the Stanford-Binet than children with black mothers and white fathers. They interpret this finding as most probably indicating the influence of the childrearing practices used by the children's mothers. The findings, while supportive of an environmental interpretation, are not very critical. Mothers of interracial children may not be representative of either the black or the white populations of the U.S. and their childrearing practices may be deviant. In addition, in the absence of parental intelligence test scores, it is not possible to eliminate the influence of assortative mating and possible genetic explanation. In this connection, the authors report that the mixed parents with white mothers had, on the average, one more year of schooling than the mixed parents with black mothers.

Conclusion: Racial Differences in Intelligence

Despite the extensive literature on racial differences in intelligence, the studies collectively or individually fail to provide critical data in support of an explanation for the obtained racial differences in intelligence test scores. In our judgment, for the reasons given earlier, none of the studies cited by Jensen as providing evidence for a genetic hypothesis provides support for that hypothesis. The only study known to us that provides a relatively direct test of the genetic hypothesis is Eyferth's study of the intelligence test scores of illegitimate children in Germany, and his study fails to support the hypothesis and does indeed provide evidence, although not definitive evidence, for an environmental interpretation of racial differences in intelligence. In addition to the lack of evidence in favor of the genetic hypothesis, there is at least some data that suggest intelligence tests may not be measures of the same construct in the black and white samples. Scarr-Salapatak's study suggests lower heritability of intelligence test scores among blacks. Crano, Kenny, and Campbell's study suggests that the ability–achievement distinction cannot be drawn in the same way in predominantly black and predominantly white samples, or alternatively, that intelligence tests may not, to the same degree, be measures of ability in black samples. Jensen's review of the literature on sex differences indicates that there is a race-by-sex interaction in test scores implying that there are characteristics of the black experience—either biological or cultural—that depress test scores in black males or, alternatively, raise them in black females. And finally, Duncan, Featherman, and Duncan's analysis of social mobility in American society implies that a model of social mobility employing test scores constructed for white samples cannot be used to explain social mobility in black samples. These data bear on an issue that has not received sufficient consideration in discussion of racial differences in intelligence test scores. It has often been implicitly assumed in such discussions that intelligence tests are measures of the same construct in different social groups. The interest in intelligence test scores stems both from a concern with the predictive validity of such test scores (i.e., the ability to predict socially relevant characteristics from knowledge of test scores) and from a concern with the construct validity of test scores as a measure of ability. Indeed, Cyril Burt defined the construct, "intelligence" (N.B. not the test score) as innate ability. And, much of the social interest surrounding test scores stems from the belief that they reflect the construct intelligence—or basic intellectual ability which is causally related to the ability to acquire the socially useful skills taught in the schools. The debate surrounding racial difference in intelligence test scores has been construed, perhaps unjustifiably, as a discussion of the reasons for racial differences in intelligence rather than a discussion of racial differences in intelligence test score. Intelligence test scores do not have the same scientific status as measures of height. For virtually all practical and theoretical purposes, measures of height are equivalent in meaning to the construct they measure. Intelligence test scores are only tenuously related to the construct they allegedly measure and it requires a rather elaborate inference to assert that they are measures of the construct. Quite apart from the difficulty of justifying that inference with respect to the white population of the U.S., we have seen that the inference may be less justifiable with respect to the black population of the U.S. This in turn implies that a discussion of racial differences in "intelligence" may not be justified because we do not yet know that such differences exist—although we do know that there are differences in test scores. And, differences in test scores imply differences in "intelligence" only if the tests are to equal degrees measures of the same construct in both groups. This may not be the case.

The debate on the reasons for differences in intelligence test scores has implicitly assumed that the practical import of the theoretical conclusion is quite different depending upon whether a genetic or environmental answer is provided. It has sometimes been assumed that a genetic answer is far more devastating or socially reprehensible than an environmental answer. Such a view may be simplistic and may be based on the erroneous view that genetic influences on phenotypic characteristics are invariant and cannot be manipulated by environmental intervention. In point of fact, environmental interventions can, in principle, change the influence of the genotype on the phenotype. Also, the environmental factors that might be responsible for inadequate development of intellectual functioning may not be readily subject to change, and may in fact be less subject to feasible environmental intervention than possible genetic influences on the development of intelligence. Perhaps, the principal applied significance of belief in the genetic explanation of racial differences in intelligence test scores is that it erroneously leads to a kind of defeatism and serves to justify among schoolteachers and others a belief in the impossibility of providing adequate educational achievement among black children. Also, belief in a genetic explanation of racial differences in intelligence may, and often has been, overgeneralized in its import and has been used by racists to justify segregation and racial discrimination. While such "policy" implications do not follow from the genetic theory—especially in view of the fact that race per se is not the major determinant of test score and the intrarace variance is larger than the variance attributable to differences between the races—the psychologist who puts forward a genetic explanation for racial differences should at very least be aware of the possible misuse of his work. We believe, in view of the lack of scientific evidence in favor of a genetic explanation, that its promulgation is not only premature but unwarranted and unwise.

Birth Order, Family Size, and Intelligence

In addition to race and social class there are two other major demographic variables relating to intelligence test scores. These are family size and birth order. It has long been known that children belonging to large families tend to score lower on tests of intelligence than children of small families (see Anastasi, 1956, for a comprehensive review of this literature). Results of studies comparing the test scores of children of different-size families are ambiguous about the reason for the lower test scores among children belonging to large families. That is, it is possible that children growing up in large families are less likely to receive parental attention and as a result are more likely to experience an environment that inadequately fosters intellectual development. However, it is also the case that such studies invariably involve interfamily comparisons and, as a result, are open to interpretations suggesting that families that are less likely to provide favorable genetic or environmental influences on intellectual development are more likely to have large families. On this latter interpretation it is not family size per se that influences intellectual development but rather the kinds of parents that elect to have large families. In this connection it should be noted that family size is inversely related to socioeconomic background. However, this is not the total explanation of the relationship between family size and intelligence because there are data suggesting that the relationship between family size and intelligence persists even after social class is controlled (Nisbet & Entwistle, 1967). However, control for social class quite clearly is not control for all the potential variables that may influence intellectual development. For example, among families of the same social class background those electing to have more children may be more capable of fostering intellectual development.

In addition to the influence of family size, there is a literature suggesting that birth order influences intellectual development. Firstborn and only children have frequently been found to be overrepresented among the intellectually eminent (Sampson, 1965). However, it was not until 1973 that systematic data became available on the influence of birth order on intelligence. Belmont and Marolla (1973) reported the results of Ravens test scores for 386,114 Dutch selective service registrants. The sample represented virtually the total male population of the Netherlands who became 19 years of age in the years between 1963 and 1966 inclusive. These data have a number of desirable characteristics for the examination of the effects of family size and birth order on intelligence. The use of data on an entire population eliminates sampling bias. The size of the sample permits one to examine the effects of birth order among different social class groups and within families of different size. And, finally, the inclusion of individuals who attained the same age at the same time (i.e., of a cohort) enables one to control for the influence of secular changes in test score associated with the year of one's birth. Figure 6.4 presents a graphic representation of the Belmont and Marolla data based on an analysis of these data by Zajonc and Markus (1975). An examination of Figure 6.4 reveals several of the more important features of these data. First, intelligence as measured by the Ravens test declines with family size. Second, within each family size children born earlier tend to score higher on the tests than children born later. Third, children born last tend to show a substantially larger drop in test score than the decline between other adjacent siblings. In fact there is some slight indication in these data that the decline between adjacent siblings decreases as birth order increases, e.g., the decline between the first and second child appears

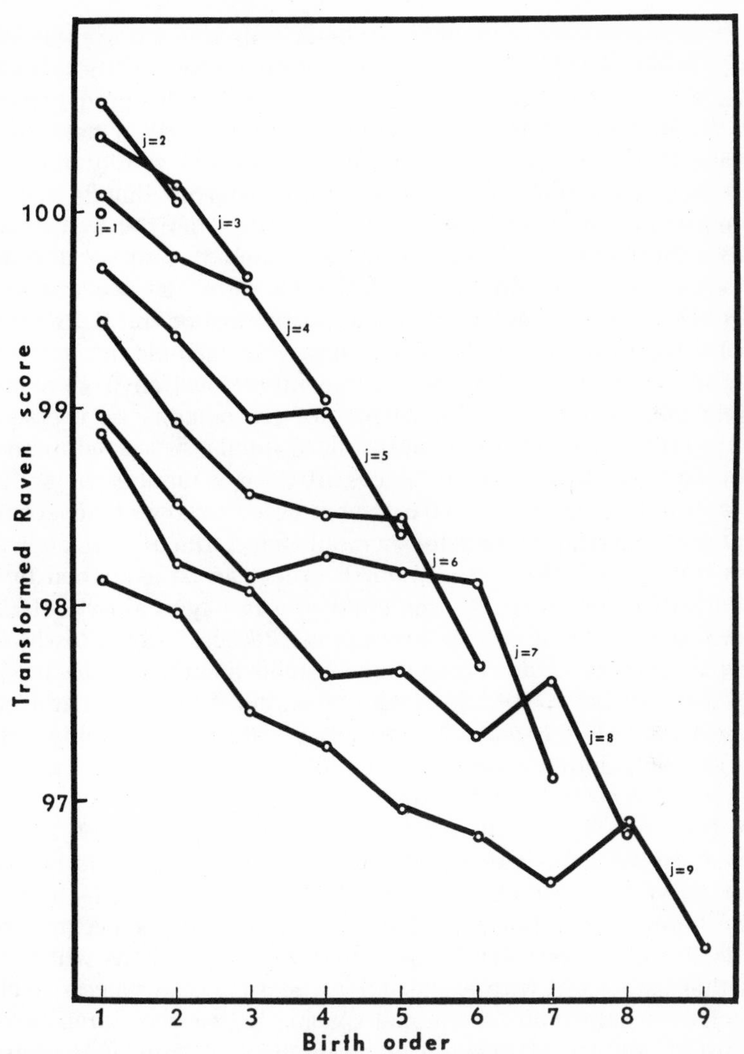

Figure 6.4. *Birth order, family size, and intelligence. [From Zajonc, R.B., & Markus, G.B. Birth order and intellectual development.* American Psychologist, 82, *75.Copyright 1975 by The American Psychological Association. Reproduced by permission.*

to be perceptibly larger than the decline between the fourth and fifth child. Thus the last child appears to occupy an anomalous position. Similarly, the only child (who is both first and last born) has a score lower than that of the first born in families with two and three children.

Zajonc and Marcus (1975) have suggested an interpretation of these data. They attempt to explain several features of these data by appeal to what they call a "confluence model." This model assumes that the intellectual level of a child is determined, in part, by the interactions among all members of the primary family unit. The intellectual level of a family, for purposes of quantifying its expected influence on intellectual development, can be derived by averaging the intellectual level of all its members including the child whose intellectual level is being considered. This simple averaging assumption explains a number of the essential features of the Belmont and Marolla data. Intellectual level decreases with increases in family size because in larger families the average intellectual environment impinging upon a child is of necessity lower, including as it does the average intellectual level of adults whose intellectual development is essentially complete and the scores of a large number of children whose intellectual growth is not completed and who in absolute terms (not age-related) have lower intelligence. The decreasing intelligence test score with increasing birth order can be explained by the same principle. Each succeeding child in a family of a particular size is born into a family whose average intellectual level is lower. For example, if parents are arbitrarily assigned an intelligence level of 100, the first-born child in a family enters a family environment whose average is 67—assuming that the child has an intelligence of zero at birth and averaging the child's intellectual level with the parents'. The second child born to this family might be born at the time that the first child has attained an intelligence of 20. The average intellectual environment for the second child would be 100 + 100 + 20 + 0 divided by 4, or 55. Thus the average intellectual level of the family is less favorable for the second child. As a general rule, the decline in average levels associated with later births will be less than that associated with earlier births because each successive child (especially where there is minimal spacing between children) adds a diminishing reduction to the average intellectual level of the family. For example, the birth of a first child, on our previous assumptions, decreases the average intellectual level of the family from 100 to 67. The birth of a second child would decrease the average level to 55. The birth of a third child to this family might occur when the second child has attained an intelligence of 20 (N.B. we assume equal spacing between the second and third and the first and second child). At this time, the first child has attained an intelligence test score of 30 (N.B. the growth of intelligence is assumed to be defined, perhaps with the exception of very early life, by a negatively accelerated curve, thus providing increasingly slower growth with age through the childhood years). The average intellectual level of the environment of the third child, on these assumptions, is derived by adding 100 + 100 + 30 + 20 + 0 divided by 5, which is 50. Note that the decline in the intellectual level of the family has decreased from the birth of the second to the birth of the third child. Note further that the decline in the average intellectual level of the family with the birth of each successive child does not invariably occur. If births are widely spaced so that older siblings have attained an intellectual competence close to that of their adult status prior to the birth of a new child, then the average level of the intellectual environment of a sibling born later may be higher than the average level of the intellectual environment impinging on a sibling born earlier. Thus the variable of spacing between children is a crucial determinant of birth order effects on intelligence, according to the confluence model.

The confluence model predicts that the only child should have the highest level of intellectual competence of any birth position. Furthermore, the model predicts that the child born last should show, on the average, a smaller decrease in intelligence relative to the intelligence of the next-to-last-born child than the decrease in intelligence between other adjacent pairs. Thus the confluence model does not explain the intelligence test scores of only children in the Belmont and Marolla data, nor does the model explain the precipitous decline of the last born child in their data. In order to explain these data, Zajonc and Marcus appeal to the notion that the opportunity to teach younger siblings fosters intellectual development. The child born last is deprived of this opportunity. And, as noted earlier, the only child is both first and last born.

The confluence model attempts to explain the influences of birth order and family size on intellectual development by appeal to the influence of interaction patterns and socialization practices for cognitive competence which emerge in different family constellations. Other types of explanations are possible. In particular, it is possible that biological factors may account for some or all of the effects of birth order. The effects of birth order (as opposed to family size) are not attributable to genetic mechanisms because siblings, within current genetic theory, are represented as having randomly selected combinations of parental genes. Thus, siblings of different birth orders should not differ in thier genetic endowments. However, the biological process of gestation and birth could, in principle, influence intellectual ability. It is possible, for example, that mothers provide an increasingly inadequate prenatal environment with successive births. Also, the variable of birth order is confounded with maternal age and older mothers may provide less adequate prenatal environments than younger mothers. The anomalous scores of the children born last in the Belmont and Marolla data might also be subject to biological interpretation. Perhaps mothers who have experienced difficult births or some biological trauma in connection with the birth of a child would be less likely to reproduce again. Therefore, children born last would represent a biased sample of children whose mothers experienced pregnancy or birth complications. In this connection Belmont, Stein, and Susser (1975) have reported that only children in the Dutch military induction sample were somewhat shorter in height than were first-born children in two-child families. In addition, only children have been found to have a higher military rejection rate (although not necessarily for health reasons) than first- or last-born children in families of two, three, or four members (L. Belmont, personal communication 1975). Explanations of the anomalous intellectual position of the only child in the Belmont and Marolla data should be accepted cautiously in view of the fact that other data exist that suggest that only children are intellectually superior to other first-born children. For example, in the large-scale survey of intelligence test scores for 11-year-olds in Scotland, only children had the highest scores (Scottish Council for Research in Education, 1949).

The confluence model can be used to explain other data. The average intelligence test scores of twins is lower than that of singletons. Also, triplets have been found to score lower than twins (Record, McKeown, & Edwards, 1970). In addition, the surviving twin of a pair in which one twin dies near birth tends not to show the intellectual deficit of several points in IQ associated with being a twin. These data suggest that the process of interaction associated with being a twin or a triplet decreases intelligence. Note that twins are born into a family that, given the confluence model, of necessity is one with a lowered intellectual environment.

Although the confluence model nicely explains the effects of birth order and family size on intelligence, the crucial test of the model has yet to be performed. The model predicts, as we have noted, that the spacing between children is a critical determinant of the birth order effect. This aspect of the model has yet to be tested. If predictions about the influence of spacing derived from the confluence model were supported, then belief in its validity would be substantially increased.

7

Conclusion: The Use of Intelligence Tests

Intelligence tests are frequently considered a positive applied contribution of psychology. Because such tests were designed to solve a practical problem and because they are widely used in applied settings, we believe it is important to discuss the use and abuse of intelligence tests.

Tests and the Public Schools

The most widespread use of intelligence tests is within the public schools. Virtually every American child is given some general test of intelligence—typically a group test—some time during his public school career. We have a number of reservations about this practice and we are not at all sure that this use of tests is justified.

It should be noted that the scores obtained and then placed in a student's record typically represent estimates of intelligence test performance obtained under less than optimal conditions. First, the scores are typically derived from group rather than individually administered tests. Second, the tests are administered on a single occasion rather than on two or three occasions briefly separated in time in order to reduce the influence of test unreliability in obtaining an estimate of test score.

In addition to the possibility that a student's test score on his school record may not be an optimal estimate of his performance on intelligence tests, it is quite likely that the meaning of the test score will be misinterpreted by many of the school personnel who will have access to the test score. We are not particularly sanguine about the possibility of training educators to the proper use of intelligence test scores. For example, many educators, apparently without awareness of the data obtained in longitudinal studies of intelligence through the school years (such as Bayley's study), believe that intelligence test scores are fixed and invariant. Also, we believe that educators are inclined to be overly rigid in their inferences from test scores to possible vocational accomplishment. For example, students with relatively low scores are occasionally dissuaded from taking college preparatory courses even where there is a reasonable possibility of completing college.

Advocates of intelligence tests and their role in education are inclined to argue that the misinterpretation of test data which may occur does not preclude the use of tests, but merely indicates that greater effort should be expended in training individuals in the proper use of tests and greater care should be taken in interpreting the results of tests. However, as a practical matter, as long as test scores are widely disseminated within a school, it is unlikely that only persons with a necessary degree of training are likely to use test scores to influence actions that may affect a student. Also, it is not invariably the case that individuals who are professionally trained to administer and interpret intelligence tests are properly equipped to do so. In this connection we have encountered many psychologists and students who received training at the graduate level in the administration and interpretation of intelligence tests in which they were misinformed about properties of intelligence tests by their instructors. For example, we have encountered countless psychologists who were told that scores on intelligence tests were normally distributed. While one could not necessarily expect students and instructors in the typical clinically oriented course in intelligence testing to be aware of Cyril Burt's work on the distribution of intelligence, one would expect them to be aware of Wechsler's work on this topic and of the information on the distribution of the test in Wechsler's test manuals.

In addition to the misinformation that occasionally exists about the basic psychometric properties of the test, we have encountered trained psychologists who are either uninformed or misinformed about basic data required for adequate test interpretation. This includes lack of information about studies dealing with stability and change in test score, and lack of information about the newer data on change in test score with age which vitiates earlier views stressing an inevitable age-related decline in test score (see Chapter 3).

In summary, we believe that test scores either in the hands of professional psychologists or educators have a great potential for misinterpretation by professionals who lack adequate knowledge of the limitations of tests and the proper inferences that may be drawn from them.

Quite apart from the potential for misinterpretation of tests is the question of their usefulness to educators in the public schools. As we have noted elsewhere, the fact that tests are able to predict school achievement is of little relevance to the schools because school achievement is known to educators. While the predictive value of intelligence tests is well documented (see Chapter 4), they have little or no diagnostic use for providing insights into specific intellectual processes. Accordingly, the tests do not provide much information that would be of specific educational relevance. Furthermore, little or no progress has been made in the design of different curricula for students of different intellectual levels. Indeed, we are not aware of any data indicating that tests of general intellectual ability would be particularly valuable as instruments for the selection of pupils who might benefit from specific curricula.

Intelligence tests can be and often are used to create student groups homogeneous in ability. Of course, such grouping could be accomplished by a teacher's observation of a student's progress in learning school material. This latter method of grouping might be advantageous if it were more sensitive to changes in pupil performance that would require a change in placement. Ability grouping by test performance, on the other hand, carries with it the chances of legitimizing caste-like assignment of children to groups, since the assignment is rationalized by reference to an assumed ability that is not presumed to be responsive to environmental intervention.

In addition to the possibility that intelligence tests may not be needed and may not provide an optimal basis for the creation of homogeneously grouped students, there is at least some indication that the kind of gross ability grouping that can be accomplished by intelligence tests may not be desirable. The available evidence suggests that ability grouping is not a particularly effective method of increasing academic achievement. Contrary to the claims often made on its behalf, there is little or no indication that students in "fast" classes are able to achieve more than they can in heterogeneously grouped classes (Borg, 1966; Findlay & Bryan, 1971). Finally, students who are grouped by ability into low groups may develop negative attitudes toward themselves and school.

Although we have reservations about ability grouping, we believe that such reservations are not incompatible with a position that favors the individualization of curriculum and instruction. Students may need to progress through a curriculum at their own rate. Also, through the introduction of "branching" procedures, different students may require a curriculum that provides different experiences at different times. One student may safely skip a section of material—another student may require intensive additional sets of practice materials in order to master some skill. However, we do not believe that individualized curricula can optimally be selected by scores on intelligence tests. An alternative strategy might be to use specific achievement tests to obtain information about student progress. This information can then be used to provide that student with an appropriate curriculum. We believe that use of achievement tests, teacher observations, or relatively specific "criterion referenced tests" designed to provide information about relatively specific skills are superior to the use of intelligence tests as a basis for the individualization of instruction.

Although it is difficult to ascertain the positive value of intelligence tests as used in the public schools, it is not difficult to suggest possible negative consequences that follow from the use of tests. Mercer (1973, 1974) has discussed what she calls latent functions of tests. One of these is what she calls the "cooling out" function. "Cooling out" refers to a use of intelligence test scores as a way that educators may justify poor academic achievement. By citing low intelligence scores as a fixed ability characteristic outside the control of the schools, an educator can attribute poor achievement not to the failure of the schools but to characteristics of students which are independent of schools. Such a use of test data is usually supported by a conceptual model in which a sharp distinction is drawn between tests of ability and achievement and in which scores on the latter are seen as the results of the causal influence of the former, i.e., ability causes achievement, rather than the converse. The citation of intelligence test scores of children growing up in poverty and of black students in this way essentially serves the conservative purpose of absolving schools and teachers for poor achievement. Yet, paradoxically, it is precisely for this group of children that the distinction between ability and achievement cannot be confidently drawn (see discussion of Crano, Kenny, and Campbell's study in Chapter 6). Clark (1965) has written eloquently about his belief that teachers should view low intelligence test scores not as a fixed limitation to the academic achievement of impoverished children but in fact as the result of an inadequate educational experience which can be remedied. He states:

> Educators, parents, and others really concerned with the human aspects of American public education should dare to look the I.Q. straight in the eye, and reject it or relegate it to the place where it belongs. The I.Q. cannot be considered sacred or even relevant in decisions about the future of the child. It should not be used to shackle children. An I.Q. so misused contributes to the wastage of human potential. The I.Q. can be a valuable educational tool within the limits of its utility, namely, as an index of what needs to be done for a particular child. The I.Q. used as the Russians use it, namely, to determine where one must start, to determine what a particular child needs to bring him up to maximum effectiveness, is a valuable educational aid. But the I.Q. as an end product or an end decision for children is criminally neglectful. The I.Q. should not be used as a basis for segregating children and for predicting—and, therefore, determining—the child's educational future [p. 129].

The quotation from Clark's book suggests that test scores themselves be the target of intervention. An alternative strategy has been suggested by Bloom (1974). Rather than attempt to change test scores, educators ought to consider whether modifications in educational practices exist that would have as their result a decrease in the relationship between intelligence test scores and academic achievement. We know that under present conditions of schooling, achievement as reflected in what is learned in school is substantially predictable by intelligence test scores. Bloom advocates the use of mastery learning procedures (Block, 1971). In mastery learning, students are required to demonstrate knowledge of a specific skill before proceeding sequentially to the next skill. Students who are not able to demonstrate mastery of that skill are not permitted to progress to the next step in the curriculum. At the beginning of a mastery learning program there are large differences in the amount of time required to reach some specified level of mastery. The differences in elapsed time are frequently on the order of 5:1—that is, the slowest 5% of students in a regular classroom may require five times as long to attain the same level of mastery attained by the fastest 5% of the students. Differences in time on task—i.e., the actual amount of time in which a student is engaged in educationally relevant activities—are somewhat less, approximately on the order of 3:1. Bloom (1974) indicates that intelligence test scores are related to the time required to reach mastery in school situations; the correlations range from .50 to .70.

If a mastery learning curriculum is developed in which assistance and additional time is provided so that slow students are able to attain mastery, according to Bloom an important change occurs in the variations in elapsed time and time on task required to attain mastery. After considerable previous mastery experiences have taken place, the variations in elapsed time decrease to the order of 3:1 and the variations in time on task decrease to an order of approximately 1.5:1. Accompanying this decrease in variations in time for learning is a decrease in the correlation between intelligence test score and time to learn. Bloom believes that mastery learning procedures provide a method for reducing the dependence of school learning on individual intellectual aptitude. As he puts it:

> What I have been trying to describe is the way in which highly *malleable* and *alterable* characteristics—the learning of each unit in a series—replace the *less malleable* and *relatively stable* characteristics represented by measures of general intelligence or specific aptitudes [1974, p. 685].

In summary, we can find few if any positive functions for routine intelligence testing in the public schools, and some possible negative functions to which tests lend themselves. And, as a result, we do not believe that tests should be routinely used in the public schools.

Tests and Selection

Since public schools accept virtually every student living in a geographic area, the use of intelligence tests in this setting is not primarily one of selection but of information about characteristics of students—although, as we have seen, the tests may be used for selection within the schools. Somewhat different issues are raised by the use of tests for the purpose of selecting individuals. Intelligence tests can be used for the purpose of selecting individuals for positions that are generally considered desirable—attendance at college, employment. There is little doubt that there are a variety of situations in addition to the obvious case of success in college where intelligence tests can be shown to have predictive validity. That is, the tests may be shown to correlate positively with criteria of success—e.g., completion of college or grade point average (see Chapter 4). In such situations the use of tests would appear to be rational and justified in that the tests would appear to permit a person charged with the responsibility of selecting others to improve his ability to select those who can succeed in that particular situation.

Despite the apparent rationality of the use of intelligence tests for selection in situations in which they have positive correlations with a criterion of success, there are a number of considerations that bear on the use of tests for this purpose.

Any evaluation of a particular selection procedure should consider the various costs (or utilities) associated with various decisions and their outcomes (Cronbach & Gleser, 1965; Wiggins, 1973). There are some outcomes that can be distinguished in a selection situation. Each of these can be expressed in terms of a probability. There is the probability of selecting someone who will succeed in that situation—called a valid positive; there is the probability of selecting someone who will not succeed—a false positive; the probability of failing to select someone who would succeed—a false negative; and the probability of failing to select someone who would fail in the situation—called a valid negative. Each of these outcomes may be conceived of as having a cost or utility associated with the outcome. The utilities associated with valid decisions are positive and the utilities associated with false decisions are negative. The cost of testing is considered a negative utility. The utility of a selection procedure can be represented in terms of the sum of the product of each of the four probabilities times the respective utilities associated with those outcomes minus the negative utility of testing. In personnel work it has been traditional to ignore the negative utility of false negatives. That is, most institutions are less concerned with the negative consequences to an individual of their failure to select him or her, and are more concerned with the negative utility of false positives because these clearly create institutional strain and are perceived as wasted effort on the part of institutions. A policy that ignores the negative utility of false negative decisions while at the same time emphasizing the negative utility of false positives has practical consequences for selection procedures. Such an assignment of utilities may influence the ways in which tests are used for selection. For any test (or battery of tests) with greater than zero predictive validity, the probability of false positive decisions may be decreased by setting a higher cutting score on the test for selection. However, this desirable decrease in the probability of false positives occurs at the price of an increase in the probability of false negatives. These probabilities are necessarily related. However, if false negative decisions are perceived as having little or no negative utility, then it is rational from the point of view of maximizing the expected utility of a selection procedure to increase the cutting point on test scores so that a smaller number of individuals is selected. This procedure is possible where the number of applicants exceeds the number of openings. And such a procedure is, in a general sense, a reasonable analysis of the institutional procedures of selection of prestige graduate schools and colleges.

The utilities of institutions may not be equivalent to the utilities of individuals. Clearly, the negative utility to an individual of not being selected for a position or institution where he is capable of successful performance is not zero. Since tests and batteries of tests of even modest predictive validity may be used to dramatically increase the probability of valid positives, if the selection ratio may be varied by setting a relatively high cutting score at the cost of increasing the number of false negatives, it is apparent that the use of tests in general (and intelligence tests in particular) can lead to significant negative social costs in many situations. Given this analysis, what alternatives exist? Tests may be used to determine a group of individuals who have a probability of being valid positives, which is acceptable to an institution. The group of individuals so selected may form a pool of individuals significantly larger than the number who can be accepted for the position. Then individuals may be selected at random to obtain the number required. Although the probability of false negatives under this procedure might not be dramatically different from one in which tests are used to select a limited number of individuals, there is a possible advantage to this procedure. Individuals with a reasonably high probability of success in a situation would be given an equal probability of selection for the position through the use of random selection procedures. Such an approach might contribute to a perception of test fairness. That is, that the tests are not being used to exclude qualified persons.

Two objections to the use of random selection procedures, as described above, might be made. First, it could be argued that such a procedure would select a group of individuals who were, on the average, less qualified. This criticism has some degree of validity. As long as test scores are monotonically related to a criterion, then it is reasonable to assume that individuals whose scores are higher on the test would perform better. However, in many situations differences in degrees of performance among a class of successful performances may not be meaningful, important, or predictable from test scores. Consider, for example, success in college. It is true that individuals whose test scores on measures of college aptitude are extremely high may be predicted to obtain a higher grade point average than individuals of somewhat lower ability score. But it is not at all clear that there would be, above a reasonably high cutting point, any significant difference in their probability of being able to complete college. The use of a high cutting score in such a situation becomes a selection for the criterion grade point average rather than the ability to do college level work. It is well to distinguish between these criteria. And, only if it is assumed that grade point average is an adequate criterion is it reasonable to set the highest possible cutting score in such a situation. It could be argued that selection for grade point average potential construes the benefits of a college education in an overly restrictive manner. Put another way, this analysis of selection is compatible with a "threshold model" of the relationship between intelligence test score and performance in academic (and other) situations. While a minimum amount of ability may be necessary to perform in that situation, higher test scores may not be related to truly meaningful and important criteria.

A second objection to the use of lower cutting scores and random selection procedures derives from the perception that such a procedure is unfair to certain individuals in that it would inevitably result in the selection of some individuals whose test scores (or, more generally, battery of scores predicting aptitude for performance) were lower than those of individuals who were not selected. Such an objection would be valid only if we believed that higher scores were necessarily more "meritorious" in some meaningful way. There exists an infatuation with numbers and subtle comparisons. From such a perspective an intelligence test score of 128 is perceived as somehow more virtuous, desirable, and meritorious than one of 125, and the individual possessing the former is to be rewarded more than the latter. Therefore, the selection of the individual with the lower score (all other things being equal) is somehow assumed to be unfair to the person with the higher score. However, this judgment is rational only if it can be shown that the difference in scores is significantly related to relevant criteria. In addition, the marginal difference in predicted criterion difference must be balanced against the costs of false negative decisions.

Our discussion of the use of intelligence tests as a basis for selection has approached the issue without considering the possible effects of group membership on prediction. A number of articles have been published dealing with the issue of test fairness from the point of view of different groups of individuals. These issues have in part derived from the argument that intelligence tests may be biased or unfair to members of minority groups. One possible source of bias in the use of tests derives from the possibility that the predictive inferences from tests are different for different kinds of individuals. We have discussed the possibility that inferences about genotype from phenotypic scores may not be equally valid for all social groups. However, with respect to the use of tests for selection, this type of possible inequality of the meaning of test scores is not of central importance. What is of central importance is the predictive accuracy of test scores for different social groups when what is predicted is some criterion of success in a situation where test scores are used as a basis for selection. One way of dealing with this issue is in terms of the equality of regression of test scores on the criterion score for individuals belonging to different social groups. In effect, an analysis of equality of regression lines deals with the issue of whether the common prediction formula will adequately represent the relationship between test score and criterion for different groups of individuals.

Cleary, Humphreys, Kendrick, and Wesman (1975) have dealt with this issue. Their analysis involves a comparison of regression equations used for the prediction of success in college from knowledge of scores on various tests of ability. The available data indicate that the correlation between test scores and grade point average in college are comparable for blacks and whites. For example, Cleary (1968) reported correlations of .47 and .47 for the Scholastic Aptitude Tests of Verbal and Mathematical Ability and college grade point average for black students. The comparable correlations for white students were .47 and .39. In the same study she derived regression equations for black students attending three different integrated colleges. She derived a regression equation from the correlations between the two scores on the Scholastic Aptitude Tests and grade point average for black students. Then, using the same tests, she derived a regression equation for white students. In order to compare the regression equations she derived predictions from them for a hypothetical black student with average scores on each of the two tests. These scores were then inserted into the regression equation for blacks and whites. This procedure, in effect, permits one to compare the predicted outcomes from test scores for "typical" black students under conditions in which a separate regression equation is derived for them and under conditions in which they are assigned values based on the white sample. This procedure was repeated in three separate colleges. In the first college, the predicted grade point average for the black student with "typical" or average scores, using the black regression equation, was 1.82. The comparable predicted score based on the white equation was 1.89. In School 2, the predicted values for black and white equations for "typical" blacks was 1.83 and 1.81, respectively. And in School 3, the comparable predictions were 1.81 and 1.98. These predictions are only marginally different. In the case of a school with a large N (over 300 black students and over 400 white students), the differences in regression equations were statistically significant. There is some slight indication that the use of a regression equation derived for white students will slightly overpredict the performance of black students. That is, black students will be predicted to perform slightly better than they actually do on the basis of their test scores. Thus, if anything, a regression analysis suggests that the use of tests gives a slight advantage to black students. On the whole, the available data indicate that there is little persuasive reason to develop separate regression equations for black and white students in selecting them for college.

Quite apart from the issue of selection of individuals of different social groups with the same regression equation, a number of recent analyses suggest that tests that are "fair" to individuals may be "unfair" to groups of individuals whose mean score on the test is lower than that of other groups (Schmidt & Hunter, 1974; Thorndike, 1971).

That is, paradoxically, a test may be fair to individuals in the sense that it provides all individuals with the same score an equal opportunity for selection, while being unfair to groups of individuals whose mean score is lower on the test. This paradox derives from the fact of errors of measurement in prediction and is likely to be exacerbated in situations in which a relatively high cutoff point is chosen on a test in order to minimize the probability of false positive selection. We can define test fairness from a group perspective in terms of the probability of a member of a group's being selected if one would succeed if one were selected. If a selection procedure is used that involves the use of a test, then the test is fair for individuals belonging to different groups if for each group the chances of being selected are a constant proportion of those who would succeed at the task if they were selected. Paradoxically, a test that is fair to individuals will not be fair to groups. Put another way, the probability of a false negative decision will be higher for groups whose mean score is lower on the test. With respect to black students whose scores on ability tests are lower than white students', a selection procedure that uses a common regression line or cutting point for both types of students will always select a smaller percentage of **potentially successful** black students than white students. It is in this sense that a test that is fair to individuals is unfair to groups. In order to remedy this unfairness, it would be necessary to use different selection criteria for black and white students and permit black students a lower qualifying score in order to be selected. Note that this procedure is not one that is based on the argument that standards should be lowered for black students as reparation for past injustice. Nor is it a proposal based on a quota system argument according to which a selection procedure is fair only if it selects individuals in proportion to their representation in the population. We have considerable sympathy with the "reparations" position and the "proportional" position which, in effect, propose quota plans independent of the proportion of individuals of a given group in the population who are "qualified." However, at a minimum, a selection procedure should provide equal opportunity for selection among different groups of candidates who are equally likely to succeed if selected and this, paradoxically, is exactly what a test that is fair to an individual will not do.

There are two difficulties with the use of test fairness for groups as a criterion for selection. First, the argument for group membership extends to any group membership. And, given the number of potential groups to which any individual may belong, a separate decision function for each group is likely to be an administrative nightmare and to lead to a situation in which each individual is likely to define his group membership in the most favorable way for selection (i.e., in terms of his membership in the group with the lowest score on the test). Second, independent of the rationale presented here for different cutting scores, the general public is likely to perceive such a selection procedure as unfair—that is, as unreasonably advantageous to a particular group and unfair to individuals.

It is obvious that, at some social cost, intelligence tests may be used as a basis for excluding individuals from being selected for desirable positions. Even where the test is related to a reasonable criterion of outcome, the use of tests as a basis for selection tends to focus attention on relatively invariant characteristics of persons that relate to their ability to perform certain tasks. Selection procedures based on tests not only presuppose invariance in the characteristics of persons but also in the characteristics of institutional positions for which individuals are being selected. Just as Bloom argues that alterations in methods of instruction might change the relationship between intellectual ability and the outcome of education, so too, in many other situations in which ability tests are used for selection, it is possible that the relationship between test scores and criteria might be altered. Such alterations might be achieved by the use of better training programs or perhaps by changing institutional procedures. While there are undoubtedly limits to the alterability of institutions, a preoccupation with the use of tests in general, and intellectual ability tests in particular, for selection tends to focus attention on invariant characteristics of individuals—including the characteristics they do not have that make them unsuited for a particular role—rather than on the characteristics of institutional practice. As a matter of general outlook, it may often be useful to ask whether changes in institutional practice might be made that would have as their consequence the increase in the proportion of individuals who would succeed in a given role within that institution if they were selected.

In summary, selection procedures involving tests carry a commitment to the status quo and are in this sense devices for exclusion and for the reinforcement of heirarchical orderings among individuals. And the argument that the orderings are based on merit (i.e., that the use of tests leads to a meritocracy) and that the imprimatur of scientific objectivity surrounds uses of test for selection should not be permitted to obscure the essentially conservative preconception (i.e., commitment to the view that individuals and institutions have fixed characteristics) on which such a use of tests is based.

Clinical Use of Intelligence Tests

Intelligence tests, ever since their introduction, have been used as a basis for assessing other characteristics of individuals. And, the use of an individually administered intelligence test is part of the standard clinical battery of assessment procedures (along with the Rorschach and the TAT). Individuals who consult a clinical psychologist for virtually any reason are likely to be given an intelligence test. This is likely to occur even when the level of intellectual ability of the individual is not a central focus of the individual's reason for consulting a psychologist. A psychologist could obtain a rough indication of an individual's intellectual ability by the use of any of a number of brief group-administered tests—e.g., the Otis Quick Scoring Test. The use of such a test as the Wechsler adult or children's test is justified on the basis that it supposedly provides greater insight into the personality characteristics of the individual taking the test. Further, in some selection situations it is the clinical rather than the psychometric features of the test that may be used to select individuals for various social roles. And, training in the administration of individual intelligence tests almost invariably is given by clinical psychologists. Thus, the assessment of intelligence through individual intelligence tests is embedded in the context of personality assessment for clinical purposes.

We have a number of reservations about this use of intelligence tests. The typical clinical course and clinically oriented manual on the use of tests (e.g., Zimmerman & Woo-Sam, 1973) provide many suggestions for the interpretation of patterns of test scores on the Wechsler tests and hypotheses about the influence of personality characteristics on individual subtests and on the inferences that may be drawn from particular responses to particular items. However, the inferences and interpretations drawn from various observations of behavior in the test situation, responses to items, and pattern of subtest scores is rarely, if ever, supported by appeal to relevant empirical evidence. Rather such inferences are presented to the neophyte psychologist ex cathedra as the result of the accumulated wisdom of the practicing psychologist. And, indeed, it is not unknown for such inferences to be taught to psychologists even where the accumulated empirical evidence has shown them to be incorrect.

Our reservations about the clinical use of intelligence tests stem initially from skepticism about the validity of clinical assessment procedures in general (Brody, 1972, chap. 8; Wiggins, 1973). First, there is little evidence that training in clinical psychology and assessment procedures enables individuals to make predictions about personality and behavior that are more valid than those made by individuals without such training (see studies and references cited in Brody, 1972; Wiggins, 1973). If this is true, it would imply that the standard battery of clinical tests (including, inter alia, individually administered intelligence tests) may be interpreted equally well (or badly) with or without clinical training. This suggests that the accumulated insight of clinical psychologists into the meaning of responses and patterns of responses to individual test items may not be correct.

There are very few studies in the literature that provide proper[1]

---

[1] We use the term "proper" because we shall indicate shortly that many of the alleged studies of clinical validity of intelligence tests do not use appropriate methodology.

predictive validity data for inferences about personality drawn from intelligence tests. One reason for the absence of such predictive validity is that studies of the predictive validities of clinical assessment procedures may consider the predictive validity of the usual battery of tests rather than any individual part of the test. However, the battery of tests taken as a whole may have low predictive validity and, by implication, the components of the battery may also have low predictive validity. For example, Marks (1961) tested the predictive validity of a clinical interview including the administration of a usual battery of clinical tests (including intelligence tests) at a child guidance clinic. The criterion was psychotherapists' personality descriptions of patients. Presumably, the therapist who dealt with a child for a period of time would be able to assess his personality. The psychologist who administered the battery of tests and interviewed the child then made personality predictions about the child (using Q sort procedures) that were validated against the therapist's subsequent personality ratings for the same characteristics. The validity coefficient (i.e., correlation) obtained for different clinical psychologists making the prediction ranged from .09 to .32 with a median value of .22. These data, which are not anomalous, suggest that predictions about personality characteristics derived from the clinical use of tests are not likely to be valid and cast doubt on the usual clinical interpretations of intelligence tests.

In addition to the use of intelligence test performance to assess general personality characteristics, clinical psychologists have frequently followed suggestions of Wechsler and attempted to relate various psychopathological conditions to particular patterns of response on the test. These have dealt with such questions as the scatter or variability among test scores and the particular subtest scores that are elevated or depressed. The effort to find such relationships has almost invariably gone unsupported in the relevant research literature. Notwithstanding their lack of empirical support, psychologists have continued to make such inferences and have been trained to interpret profile and scatter among test scores. Matarazzo (1972), who is basically sympathetic to the clinical use of the Wechsler tests, states in the definitive fifth edition of the manual to the Wechsler tests, "Alas, hundreds upon hundreds of studies on the use of profile, pattern, or scatter analysis with the Wechsler scales conducted between 1940 and 1970 failed to produce reliable evidence that such a search would be fruitful [1972, pp. 429, 430]." His book may be consulted for references to the voluminous literature on this issue.

In addition to the difficulty of replicating and finding patterns of relationship between characteristics of intelligence test response and clinically defined psychopathological groups, the methodology used in much of the research related to this question is not appropriate for the determination of the predictive validity for personality or psychopathology of patterns of intelligence test scores. We can illustrate this point by presenting Matarazzo's summary of studies of the relationship between sociopathy and performance on the Wechsler tests. Table 7.1 presents his summary of the literature on this issue. An examination of Table 7.1 indicates that in several studies, sociopaths have been found to score higher on performance IQ than on

TABLE 7.1
Wechsler Verbal and Performance IQ for Samples of Variously Defined Adolescent and Adult Sociopaths[a]

Wechsler scale	VIQ	PIQ	FSIQ	Investigators
W-B	82.0	94.0	87.0	Weider, Levi, and Risch (1943)
	99.4	101.7	—	Strother (1944)
	76.2	80.4	76.5	Franklin (1945)
	83.8	94.5	87.6	Durea and Taylor (1948)
	82.0	98.0	89.0	Altus and Clark (1949)
	88.6	97.2	92.3	Glueck and Glueck (1950)
	80.8	86.2	83.6	Diller (1952)
	90.1	100.8	94.8	Bernstein and Corsini (1953)
	90.2	99.9	93.9	
	101.1	101.9	101.8	Walters (1953)
	82.1	89.1	84.4	
	93.6	98.5	95.2	Vane and Eisen (1954)
	87.3	99.7	92.5	Blank (1958)
	96.3	98.3	—	Foster (1959)
	104.0	104.1	104.7	Field (1960b)
	95.6	101.7	98.7	Fisher (1961)
	83.0	98.9	89.8	
	83.7	87.8	84.4	
	86.8	96.2	90.5	Manne, Kandel, and Rosenthal (1962)
WISC	87.0	92.4	88.4	Richardson and Surko (1956)
WAIS	93.8	98.3	95.4	Wechsler (1955 *Manual*, p. 21)
	98.0	102.0	99.7	Graham and Kamona (1958)
	90.7	91.6	90.5	Panton (1960)
	77.2	78.5	76.4	
Mixed	90.1	98.3	93.5	DeStephens (1953)
	86.7	91.2	87.9	
	97.6	104.0	100.1	Wiens, Matarazzo, and Gaver (1959)
	90.9	98.0	94.1	Prentice and Kelly (1963)
	89.4	95.4	91.8	

[a]Based on Matarazzo, J.D. Wechsler's measurement and appraisal of adult intelligence, 5th ed. Baltimore: Williams & Wilkins © 1972.

verbal IQ. (The table omits some unpublished data that failed to find this pattern.) It should be noted that the differences are not always large and on occasion clearly are negligible. Furthermore, the differences reported hold with respect to group means. Clearly, substantial numbers of individuals classified as sociopaths will score higher on verbal IQ than on performance IQ. Thus, prediction about pattern of IQ score for an individual from knowledge that he is a sociopath is quite likely to be fraught with error. Matarazzo is fully aware of these difficulties. Nevertheless he asserts, "the *trend* so obvious in this table for a higher PIQ over VIQ in these vastly different and *crudely* composed groups of so-called sociopaths is too compelling for the serious student of personality-intellectual behavior relationship to dismiss [1972, p. 434.]."

One issue Matarazzo does not discuss is the direction of inference about personality that is permitted by data of the sort reported in Table 7.1. The table presents data about a pattern of response-given information about criterion group membership. Such data, strictly speaking, do not provide evidence for the predictive validity of a test. That is, the data permit the derivation of a conditional probability of test response-given criterion group membership. The appropriate conditional probability required for determining the usefulness of the test for clinical assessment is the probability of clinical group membership given particular test response. And these two conditional probabilities are not, in general, equal. If one wishes to use a test to make inferences about an individual, knowledge of the pattern of test scores of individuals belonging to various criterion groups (which is the usual data available in the literature) is useless. And, since conditions like sociopathy occur infrequently in the population, it is virtually certain that inferences about sociopaths from patterns of verbal and performance IQ will be wrong. That is, the probability of a higher performance than verbal IQ is approximately .5. The probability that an individual who has a higher performance than verbal IQ will be classified as a sociopath may be negligibly different from the probability that any individual selected at random will be classified as a sociopath. Hence, knowledge of pattern test score is worthless for predicting sociopathy.

More generally, this analysis suggests that the failure to distinguish between the two types of conditional probabilities—p(C/R) = probability of criterion-given response that is relevant to test validity, and p(R/C) = probability of response-given criterion that is not relevant to test validity—may lead to an overinterpretation of the psychological significance of test scores. Clinically defined groups may differ in many ways from "normal" individuals and thus may exhibit many responses different from those exhibited by "normal" individuals. However, "normal" individuals may exhibit the same responses and not be a member of the particular clinically defined group. Hence, inferences about psychopathology from test responses may not be correct.

Matarazzo presents an example of clinical inferences from an individual intelligence test that influenced a selection decision (1972, pp. 501–503). He describes an examination of a 21-year-old applicant for a position as a highway patrolman who, on the basis of Matarazzo's evaluation, was rejected for the position. The principal reason for the recommended rejection was the clinical evaluation of performance on the intelligence test. Matarazzo remarks that other than difficulty with the performance section of the IQ test, there was an absence of clinical and test indices of psychopathology. He states, "The most striking finding was the applicant's frank, clinically apparent confusion on four of the five performance subtests." This confusion led Matarazzo to suspect drug usage. Matarazzo goes on to assert, "A global index of this applicant's disability in intellectual functioning (even with etiology unknown) is the 17-point decrement in PIQ (86) relative to his own VIQ (103) obtained on the same sitting . . . such a 17-point V–P differential occurred very infrequently (only six times in a 100) in the normal WAIS standardization samples. The present writer's clinical experience also reveals that persons (normals by other criteria) scoring in the middle range of IQ almost never show such a large V–P differential [1972, pp. 501, 503]."

Matarazzo's analysis of this protocol is based on a number of generalizations:

1. Young people, in the early 1970s, who became confused during the performance part of an intelligence test being used for selection for a job, may be drug users.
2. A relatively large discrepancy between performance and verbal IQ in the middle range of IQ scores implies psychopathology.
3. Individuals who have psychopathological characteristics (of whatever kind?) are likely to be inadequate highway patrolmen. And by implication,
4. Individuals who show a "global disability in intellectual functioning" by virtue of a relatively large verbal–performance IQ discrepancy are unlikely to be successful highway patrolmen.

What is remarkable about these inferences is that there is little or no evidence for any of them. In fact, Inference 2 is specifically contradicted in the available empirical literature and the lack of relationship between discrepancy and psychopathology is implicitly recognized by Matarazzo in his summary of the available literature on the subject. It is apparent that a sophisticated and knowledgeable clinician like Matarazzo is prepared to make inferences on the basis of patterns of response on individually administered intelligence tests which are without firm empirical foundation. Furthermore, in our judgment, decisions involving selections based on such inferences **in the absence of data providing evidence for the validity of such inferences in terms of correlations between patterns of test response and relevant criterion measures of performance** are entirely unwarranted. In our judgment, there is little or no evidence at present that patterns of response to individually administered intelligence tests provide useful information about individuals. This in turn suggests that much of the training in the use and interpretation of intelligence tests provided to psychologists may be misdirected in its emphasis. We would prefer greater emphasis on the psychometric characteristics of the tests and acquaintance with the relevant empirical literature surrounding the test. Thus, the potential for unwarranted uses of the tests extends from improper inferences and exclusions based on overall test scores to improper inferences and exclusion based on interpretation of the responses that led to an overall global score.

A Concluding Comment

We have been rather negative about many standard uses of intelligence tests. Our negative comments do not derive principally from a belief in the lack of validity of existing tests. On the whole, psychologists have been rather successful in devising tests that measure the kinds of abstract intellectual skills that are useful for success in school. However, this tour de force must not, in our judgment, lead to an overreliance on the importance of this skill as a basis for the creation of a social order. It is well to reiterate that skill at answering questions on intelligence tests is not equivalent to "truth, beauty, and goodness" and is only one of the many kinds of ways in which human beings may differ and may excel.

References

Anastasi, A. Intelligence and family size. *Psychological Bulletin*, 1956, *53*, 187–209.

Anderson, L. D. The predictive value of infant tests in relation to intelligence at five years. *Child Development*, 1939, *10*, 203–212.

Anderson, V. V. *Psychiatry in industry*. New York: Harper, 1929.

Ball, R. S. The predictability of occupational level from intelligence. *Journal of Consulting Psychology*, 1938, *2*, 184–186.

Belmont, L., & Marolla, F. A. Birth order, family size, and intelligence. *Science*, 1973, *182*, 1096–1101.

Belmont, L., Stein, Z. A., & Susser, M. W. Comparison of associations of birth order with intelligence test score and height. *Nature*, 1975, *255*, 54–56.

Benson, V. E. The intelligence and later scholastic success of sixth grade pupils. *School and Society*, 1942, *55*, 163–167.

Bills, M. A. Relation of mental alertness test score to position and permanency in company. *Journal of Applied Psychology*, 1923, *7*, 154–156.

Binet, A. Review of C. Spearman: The proof and measurement of association between two things; General intelligence objectively determined and measured. *Annee Psychologie*, 1905, *11*, 623–624.

Binet, A., & Henri, V. La psychologie individuelle. *Annee Psychologie*, 1896, *2*, 411–465.

Binet, A., & Simon, T. Sur la necessité d'établir un diagnostic scientifique des états inferieurs de l'intelligence. *Annee Psychologie*, 1905, *11*, 163–169. (a)

Binet, A., & Simon, T. Méthodes nouvelles pour le diagnostic du niveau intellectuel des anormaux. *Annee Psychologie*, 1905, *11*, 191–244. (b)

Binet, A., & Simon, T. Applications des méthodes nouvelles au diagnostic du niveau intellectual chez des enfants normaux et anormaux d'hospice et d'école primaire. *Annee Psychologie*, 1905, *11*, 245–336. (c)

Binet, A., & Simon, T. Le développement de l'intelligence chez les enfants. *Annee Psychologie*, 1908, *14*, 1–94.

Birch, H. G., & Gussow, J. D. *Disadvantaged children: Health, nutrition, and school failure.* New York: Harcourt, 1970.

Birch, H. G., Piñeiro, C., Alcalde, E., Toca, T., & Cravioto, J. Kwashiorkor in early childhood and intelligence at school age. *Pediatric Research,* 1971, *5,* 579–584.

Blalock, H. M. Jr. *Causal inferences in nonexperimental research.* Chapel Hill: University of North Carolina Press, 1964.

Block, J. H. (Ed.). *Mastery learning: Theory and practice.* New York: Holt, Rinehart, & Winston, 1971.

Bloom, B. S. *Stability and change in human characteristics.* New York: Wiley, 1964.

Bloom, B. S. Time and learning. *American Psychologist,* 1974, *29,* 682–688.

Bloom, B. S., & Broder, L. *Problem-solving processes of college students.* Chicago: University of Chicago Press, 1950.

Blum, M. L., & Candee, B. The selection of department store packers and wrappers with the aid of certain psychological tests. *Journal of Applied Psychology,* 1941, *25,* 78–85.

Borg, W. R. *Ability grouping in the public schools* (2nd ed.). Madison, Wisconsin: Denbar Educational Research Services, 1966.

Brody, N. *Personality: Research and theory.* New York: Academic Press, 1972.

Broman, S. H., Nichols, P. L., & Kennedy, W. A. *Preschool IQ prenatal and early developmental correlates.* Hillsdale, New Jersey: Lawrence Erlbaum Associates, 1975.

Bronfenbrenner, U. Is early intervention effective? Some studies of early education in familial and extrafamilial settings. In A. Montagu (Ed.), *Race and IQ.* New York: Oxford University Press, 1975.

Burks, B. S. The relative influence of nature and nurture upon mental development. A comparative study of foster-parent–foster-child resemblance and true-parent–true-child resemblance. *Yearbook of the National Society for the Study of Education,* 1928, *27,* 219–316.

Burt, C. *The distribution and relations of educational abilities.* London: King, 1917.

Burt, C. Is intelligence distributed normally? *The British Journal of Statistical Psychology,* 1963, *16,* 175–190.

Caldwell, B., & Drachman, R. H. Comparability of three methods of assessing the developmental level of young infants. *Pediatrics,* 1964, *34,* 51–57.

Caldwell, J., Schrader, D. R., Michael, W. B., & Meyers, C. E. Structure-of-intellect measures and other tests as predictors of success in tenth-grade modern geometry. *Educational and Psychological Measurement,* 1970, *30,* 437–441.

Cameron, J., Livson, N., & Bayley, N. Infant socializations and their relationship to mature intelligence. *Science,* 1967, *157,* 331–333.

Cattell, J. McK. Mental test and measurements. *Mind,* 1890, *15,* 373–381.

Cattell, R. B. Theory of fluid and crystallized intelligence: A critical experiment. *Journal of Educational Psychology,* 1963, *54,* 1–22.

Cattell, R. B. *Abilities: Their structure, growth, and action.* Boston: Houghton Mifflin Co., 1971.

Cattell, R. B., Blewett, D. B., & Belloff, J. R. The inheritance of personality. A multiple variance analysis determination of approximate nature-nurture ratios for primary personality factors in Q-data. *American Journal of Human Genetics,* 1955, 7, 122–146.

Cattell, R. B., Stice, G. F., & Kristy, N. F. A first approximation to nature-nurture ratios for eleven personality factors in objective tests. *Journal of Abnormal and Social Psychology*, 1957, *54*, 143–159.

Champakam, S., Srikantia, S. G., & Gopalan, C. Kwashiorkor and mental development. *American Journal of Clinical Nutrition*, 1968, *21*, 844–852.

Clark, K. B. *Dark ghetto. Dilemmas of social power.* New York: Harper & Row, 1965.

Cleary, T. A. Test bias: Prediction of grades of Negro and white students in integrated colleges. *Journal of Educational Measurement*, 1968, *5*, 115–124.

Cleary, T. A., Humphreys, L. G., Kendrick, S. A., & Wesman, A. Educational uses of tests with disadvantaged students. *American Psychologist*, 1975, *30*, 15–41.

Coleman, J. S., *et al. Equality of Educational Opportunity.* Washington, D. C.: U.S. Office of Education, 1966.

Crano, W. D., Kenny, J., & Campbell, D. T. Does intelligence cause achievement? A cross-lagged panel analysis. *Journal of Educational Psychology*, 1972, *63*, 258–275.

Cronbach, L. J. *Essentials of psychological testing* (3rd ed.). New York: Harper & Row, 1970. (a)

Cronbach, L. J. Test validation. In R. L. Throndike (Ed.), *Educational measurement.* Washington, D. C.: American Council on Education, 1970. (b)

Cronbach, L. J., & Gleser, G. C. *Psychological tests and personnel decisions* (2nd ed.). Urbana: University of Illinois Press, 1965.

Dasen, P. R. The development of conservation in aboriginal children: A replication study. *International Journal of Psychology*, 1972, 7, 75–85.

Davenport, R. K. Implications of military selection and classification in relation to universal military training. *Journal of Negro Education*, 1946, *15*, 585–594.

DeLemos, M. M. The development of conservation in aboriginal children. *International Journal of Psychology*, 1969, *4*, 255–269.

Dreger, R. M., & Miller, K. S. Comparative psychological studies of Negroes and whites in the United States: 1959–1965. *Psychological Bulletin Monograph Supplement*, 1968, *70*, Pt. 2, 3.

Drillien, C. M. *The growth and development of the prematurely born infant.* Baltimore: Williams & Wilkens, 1964.

Dudek, S. Z., Lester, E. P., Goldberg, J. S., & Dyer, B. B. Relationship of Piaget measures to standard intelligence and motor scales. *Perceptual and Motor Skills*, 1964, *28*, 351–362.

Duncan, O. D. Path analysis: Sociological examples. *American Journal of Sociology*, 1966, *72*, 1–16.

Duncan, O. D. Inheritance of poverty or inheritance of race? In D. P. Moynihan (Ed.), *On understanding poverty.* New York: Basic Books, 1968.

Duncan, O. D., Featherman, D. L., & Duncan, B. *Socioeconomic background and achievement.* New York: Seminar Press, 1972.

Dunham, J. L., Guilford, J. P., & Hoepfner, R. Abilities related to classes and the learning of concepts. Reports from the Psychological Laboratory, University of Southern California, No. 39, 1966.

Eisdorfer, C., & Wilkie, F. Intellectual changes with advancing age. In Jarvik, L. F., Eisdorfer, C., & Blum, J. E. *Intellectual functioning in adults.* New York: Springer, 1973.

Erlenmeyer-Kimling, L., & Jarvik, L. F. Genetics and intelligence: A review. *Science*, 1963, *142*, 1477–1479.

Escalona, S. K., & Corman, H. Albert Einstein scales of sensorimotor development: Object permanence. Unpublished manuscript.

Evans, D. E., Moodie, A. D., & Hansen, J. D. L. Kwashiorkor and intellectual development. *South African Medical Journal*, 1971, *45*, 1413–1426.

Eyferth, K. Lerstungen verschiedener Gruppen von Besatzunfskindern in Hamburg-Wechsler Intelligenztest für Kinder (HAWK). *Archiv für die Gesante Psychologie*, 1961, *113*, 222–241.

Findley, W. G., & Bryan, M. M. *Ability grouping, 1970: Status, impact, and alternatives.* Athens: University of Georgia, Center for Educational Improvement, 1971.

French, J. W. Kit of reference tests for cognitive factors. Princeton, New Jersey: Educational Testing Service, 1963.

Furby, L. Interpreting regression toward the mean in developmental research. *Developmental Psychology*, 1973, *8*, 172–179.

Galton, F. *Hereditary genius.* London: MacMillan, 1869.

Galton, F. *Inquiries into human faculty and its development.* London: MacMillan, 1883.

Getzels, J. W., & Jackson, P. W. *Creativity and intelligence.* New York: Wiley, 1962.

Golden, M., & Birns, B. Social class, intelligence, and cognitive style in infancy. *Child Development*, 1971, *42*, 2114–2116.

Gottfried, A. W., & Brody, N. Interrelationships between and correlates of psychometric and Piagetian scales of sensorimotor intelligence. *Developmental Psychology*, 1975, *11*, 379–387.

Grinsted, A. D. *Studies in gross bodily movement.* Unpublished Ph.D. dissertation, Louisiana State University, 1939.

Guilford, J. P. Zero intercorrelations among tests of intellectual abilities. *Psychological Bulletin*, 1964, *61*, 401–404.

Guilford, J. P. *The nature of human intelligence.* New York: McGraw-Hill, 1967.

Guilford, J. P. Some misconceptions of factors. *Psychological Bulletin*, 1972, *77*, 392–396.

Guilford, J. P., & Hoepfner, R. *The analysis of intelligence.* New York: McGraw-Hill, 1971.

Guilford, J. P., Hoepfner, R., & Peterson, H. Predicting achievement in ninth-grade mathematics. From measures of intellectual-aptitude factors. *Educational and Psychological Measurement*, 1965, *25*, 659–682.

Härnquist, K. Relative changes in intelligence from 13–18. I. Background and methodology. *Scandinavian Journal of Psychology*, 1968, *9*, 50–64. (a)

Härnquist, K. Relative changes in intelligence from 13–18. II. Results. *Scandinavian Journal of Psychology*, 1968, *9*, 65–82. (b)

Harrell, R. F., Woodyard, E. R., & Gates, A. I. The influence of vitamin supplementation of the diets of pregnant and lactating women on the intelligence of their offspring. *Metabolism*, 1956, *5*, 555–562.

Harrell, T. W., & Harrell, M. S. Army General Classification Test scores for civilian occupations. *Educational and Psychological Measurement*, 1945, *5*, 229–239.

Hartson, L. D., & Sprow, A. J. Value of intelligence quotients obtained in secondary school for predicting college scholarship. *Educational and Psychological Measurement*, 1941, *1*, 387–398.

Hauser, R. M., & Dickinson, P. J. Inequality on occupational status and income. *American Educational Research Journal,* 1974, *11,* 161–168.

Hay, E. N. Predicting success in machine bookkeeping. *Journal of Applied Psychology,* 1943, *27,* 483–493.

Hebb, D. O. Intelligence in man after large removals of cerebral tissue: Report of four left frontal lobe cases. *Journal of General Psychology,* 1939, *21,* 73–87.

Hebb, D. O. The effect of early and late brain injury upon test scores, and the nature of normal adult intelligence. *Proceedings of the American Philosophical Society,* 1942, *85,* 275–292.

Heber, R., & Garber, H. An experiment in the prevention of cultural–familial mental retardation. Paper presented at the Second Congress of the International Association for the Scientific Study of Mental Deficiency, Warsaw, Poland, 1970.

Herrman, L., & Hogben, L. The intellectual resemblance of twins. *Proceedings of the Royal Society of Edinburgh,* 1933, *53,* 105–129.

Herrnstein, R. J. *I Q in the meritocracy.* Boston: Little, Brown & Co., 1973.

Hertzig, M. E., Birch, H. G., Richardson, S. A., & Tizard, J. Intellectual levels of school children severely malnourished during the first two years of life. *Pediatrics,* 1972, *49,* 814–824.

Hindley, C. B. Stability and change in abilities up to five years: Group trends. *Journal of Child Psychology and Psychiatry,* 1965, *6,* 85–99.

Hinkelman, E. A. Relationship of intelligence to elementary school achievement. *Educational Administration and Supervision,* 1955, *41,* 176–179.

Holland, J. L., & Richards, J. M. Jr. Academic and nonacademic accomplishment: Correlated or uncorrelated? *Journal of Educational Psychology,* 1965, *56,* 165–174.

Holly, K. A., & Michael, W. B. The relationship of structure-of-intellect factor abilities to performance in high school modern algebra. *Educational and Psychological Measurement,* 1972, *32,* 447–450.

Honzik, M. P. Developmental studies of parent–child resemblance in intelligence. *Child Development,* 1957, *28,* 215–218.

Horn, J. L. Organization of abilities and the development of intelligence. *Psychological Review,* 1968, *75,* 242–259.

Horn, J. L. Organization of data on life-span development of human abilities. In L. R. Goulet & P. Baltes (Eds.), *Life-span developmental psychology. Research and theory.* New York: Academic Press, 1970.

Horn, J. L., & Cattell, R. B. Refinement and test of the theory of fluid and crystallized ability intelligences. *Journal of Educational Psychology,* 1966, *57,* 253–270.

Horn, J. L., & Cattell, R. B. Age differences in fluid and crystallized intelligence. *Acta Psychologica,* 1967, *26,* 107–129.

Humphreys, L. G. Critique of Cattell: Theory of fluid and crystallized intelligence: A critical experiment. *Journal of Educational Psychology,* 1967, *58,* 120–136.

Hunt, J. McV. *Intelligence and experience.* New York: Ronald, 1961.

Husén, T. Intra-pair similarities in the school achievements of twins. *Scandinavian Journal of Psychology,* 1963, *4* 108–114.

Jacobsen, L. I., Berger, S. M., Bergman, R. L., Millham, J., & Greeson, L. E. Effects of age, sex, systematic conceptual learning sets and programmed social interaction on the intellectual and conceptual development of preschool children from poverty backgrounds. *Child Development*, 1971, *42*, 1399–1415.

Jarvik, L. F., Eisdorfer, C., & Blum, J. E. *Intellectual functioning in adults.* New York: Springer, 1973.

Jencks, C. *Inequality: A reassessment of the effect of family and schooling in America.* New York: Basic Books, 1972.

Jensen, A. R. How much can we boost IQ and scholastic achievement? *Harvard Educational Review*, 1969, *39*, 1–123. (a)

Jensen, A. R. Hierarchical theories of mental ability. In B. Dockrell (Ed.), *On intelligence.* London: Methuen, 1969. (b)

Jensen, A. R. The race *x* sex *x* ability interaction. In R. Cancro (Ed.), *Contributions to intelligence.* New York: Greene & Stratton, 1971.

Jensen, A. R. *Educability and group differences.* New York: Harper & Row, 1973.

Jensen, A. R. Cumulative deficit: A testable hypothesis. *Developmental Psychology*, 1974, *10*, 996–1019.

Jinks, J. L., & Fulker, D. W. Comparison of the biometrical, genetical, MAVA, and classical approaches to the analysis of human behavior. *Psychological Bulletin*, 1970, *73*, 311–349.

Jones, H. E., & Bayley, N. The Berkeley growth study. *Child Development*, 1941, *12*, 167–173.

José, J., & Cody, J. J. Teacher–pupil interaction as it related to attempted changes in teacher expectancy of academic ability and achievement. *American Educational Research Journal*, 1971, *8*, 39–40.

Juel-Nielsen, N. Individual and environment: A psychiatric-psychological investigation of monozygotic twins reared apart. *Acta Psychiatrica et Neurologica Scandinavia*, Monograph Supplement 1965, *183*.

Kamin, L. J. *The science and politics of I.Q.* Potomac, Maryland: Lawrence Erlbaum Associates, 1974.

Kenagy, H. G., & Yockum, C. E. *The selection and training of salesmen.* New York: McGraw-Hill, 1925.

Kennedy, W. A., VandeReit, V., & White, J. C. Jr. A normative sample of intelligence and achievement of Negro elementary school children in the Southeastern United States. *Monographs of the Society for Research in Child Development*, 1963, *28*, No. 6.

King, W. L., & Seegmiller, B. Performance of 14 to 22 month-old Black, firstborn male infants on two tests of cognitive development: The Bayley scales and the Infant Psychological Development scale. *Developmental Psychology*, 1973, *8*, 317–326.

Knobloch, H., & Pasamanick, B. An evaluation of the consistency and predictive value of the 40-week Gesell developmental schedule. *Psychiatric Research Reports of the American Psychiatric Association*, 1960, *13*, 10–13.

Knobloch, H., Rider, Harper, P., & Pasamanick, B. Neuropsychiatric sequelae of prematurity: A longitudinal study. *Journal of the American Medical Association*, 1956, *161*, 581–585.

Kogan, N., & Pankove, E. Long-term predictive validity of divergent-thinking tests: Some negative evidence. *Journal of Educational Psychology*, 1974, *66*, 802–810.

Lavin, D. E. *The prediction of academic performance: A theoretical analysis and review of research.* New York: Russell Sage Foundation, 1965.

Leahy, A. M. Nature–nurture and intelligence. *Genetic Psychology Monographs,* 1935, *17,* 236–308.

Lesser, G. S., Fifer, G., & Clark, D. H. Mental abilities of children from different social-class and cultural groups. *Monographs of the Society for Research in Child Development,* 1965, *30,* No. 4.

Lewis, M. Individual differences in the measurement of early cognitive growth. In J. Hellmuth (Ed.), *Exceptional infant,* Vol. 2. New York: Brunner-Mazel, 1971.

Lewis, M., & McGurk, H. The evaluation of infant intelligence scores—true or false? *Science,* 1972, *178,* 1174–1177.

Loehlin, J. C., Lindzey, G., & Spuhler, J. N. *Race differences in intelligence.* San Francisco: Freeman, 1975

Loehlin, J. C., Vandenberg, S. G., & Osborne, R. T. Blood group genes and Negro–white ability differences. *Behavior Genetics,* 1973, *3,* 263–270.

Maccoby, E. E., Dowley, E. M., Hagen, J. W., & Degerman, R. Activity level and intellectual functioning in normal preschool children. *Child Development,* 1965, *36,* 761–770.

MacKinnon, D. W. The personality correlates of creativity: A study of American architects. In C. S. Nielsen (Ed.), *Proceedings of the XIV International Congress of Applied Psychology.* Copenhagen: Munksgaard, 1962.

MacKinnon, D. W. Creativity of architects. In C. W. Taylor (Ed.), *Widening horizons in creativity.* New York: Wiley, 1964.

Marks, P. A. An assessment of the diagnostic process in a child guidance setting. *Psychological Monographs,* 1961, *75* (Whole No. 507).

Marolla, F. A. *Intelligence and demographic variables in a 19-year-old cohort in the Netherlands.* Unpublished Ph.D. dissertation, Graduate Faculty, New School for Social Research, 1973.

Marshall, M. V. What intelligence quotient is necessary to success? *Journal of Higher Education,* 1943, *14,* 99–100.

Matarazzo, J. D. *Wechsler's measurement and appraisal of adult intelligence* (5th ed.). Baltimore: Williams & Wilkins, 1972.

Matarazzo, J. D., Allen, B. V., Saslow, G., & Wiens, A. N. Characteristics of successful policemen and firemen applicants. *Journal of Applied Psychology,* 1964, *48,* 123–133.

Mayeske, G. W., *et al. A study of the achievement of our nation's students.* DHEW Publication No. (OE) 72-131. Washington, D.C.: U.S. Government Printing Office, 1973.

McCall, R. B., Hogarty, P. S., & Hurlburt, N. Transitions in infant sensorimotor development and the prediction of childhood I.Q. *American Psychologist,* 1972, *27,* 328–348.

McClelland, D. C. Testing for competence rather than for "intelligence." *American Psychologist,* 1973, *28,* 1–14.

McNemar, Q. *The revision of the Stanford-Binet scale: An analysis of the standardization data.* Boston: Houghton-Mifflin, 1942.

Mehrota, S. N., & Maxwell, J. The intelligence of twins. A comparative study of eleven-year-old twins. *Population Studies,* 1950, *3* 295–302.

Mercer, J. R. *Labeling the mentally retarded.* Berkeley and Los Angeles: University of California Press, 1973.

Mercer, J. R. Latent functions of intelligence testing in the public schools. In L.

P. Miller (Ed.), *The testing of black students: A symposium.* Englewood Cliffs, New Jersey: Prentice-Hall, 1974.

Miner, J. B. *Intelligence in the United States.* New York: Springer, 1957.

Moore, T. Language and intelligence: A longitudinal study of the first eight years. *Human Development,* 1967, *10,* 88–106.

Morsh, J. E., & Wilder, E. W. Identifying the effective instructor: A review of quantitative studies, 1900–1952. *USAF Personnel Training Research Center Research Bulletin,* 1954, No. AFPT RC-TR 54-44.

Munsinger, H. The adopted child's I.Q.: A critical review. *Psychological Bulletin,* 1975, *82,* 623–659. (a)

Munsinger, H. Children's resemblance to their biological and adopting parents in two ethnic groups. *Behavior Genetics,* 1975, *5,* 239–254. (b)

Newman, H. H., Freeman, F. N., & Holzinger, K. J. *Twins: A study of heredity and environment.* Chicago: University of Chicago Press, 1937.

Nichols, P. L. *The effects of heredity and environment on intelligence test performance in 4 and 7 year white and Negro sibling pairs.* Unpublished Ph.D. Dissertation, University of Minnesota. Ann Arbor, Michigan: University Microfilm, 1970, No. 71-18, 874.

Nisbet, J. D., & Entwistle, N. J. Intelligence and family size, 1949-1965. *British Journal of Educational Psychology,* 1967, *37,* 188–193.

Osborne, R. T., & Gregor, A. J. Racial differences in heritability estimates for tests of spatial ability. *Perceptual and Motor Skills,* 1968, *27,* 735–739.

Otis, A. S. The selection of mill workers by mental tests. *Journal of Applied Psychology,* 1920, *4,* 339–341.

Owens, W. A. Age and mental abilities: A second adult follow-up. *Journal of Educational Psychology,* 1966, *57,* 311–325.

Pasamanick, B., & Knoblock, H. Retrospective studies of the epidemiology of reproductive casualty, old and new. *Merrill-Palmer Quarterly,* 1966, *12,* 7–26.

Pinneau, S. R. *Changes in intelligence quotient: Infancy to maturity.* Boston: Houghton-Mifflin, 1961.

Pond, M., & Bills, M. A. Intelligence and clerical jobs. Two studies of relation of test score to job held. *Personnel Journal,* 1933, *12,* 41–56.

Record, R. G., McKeown, T., & Edwards, J. H. An investigation of the differences in measured intelligence between twins and single births. *Annals of the Human Genetic Society,* 1970, *34,* 11–20.

Reimanis, G., & Green, R. F. Imminence of death and intellectual decrement in the aging. *Developmental Psychology,* 1971, *5,* 270–272.

Roberts, J. A. F., & Sedgley, E. Intelligence testing of full-term and premature children by repeated assessments. In C. Banks & P. L. Broadhurst (Eds.), *Studies in psychology.* New York: Barnes & Noble, 1966.

Roe, A. A psychological study of eminent psychologists and anthropologists, and a comparison with biological and physical scientists. *Psychological Monographs General and Applied,* 1953, *67,* No. 352.

Rosenthal, R., & Jacobsen, L. *Pygmallion in the classroom: Teacher expectation and pupils' intellectual development.* New York: Holt, Rinehart, & Winston, 1968.

Rush, D., Stein, Z., Christakis, G., & Susser, M. The prenatal project: The first twenty months of operation. In M. Winick (Ed.), *Proceedings of the symposium on nutrition and fetal development.* New York: Wiley, 1974.

Rutter, M. Psychological development—predictions from infancy. *Journal of Child Psychology and Psychiatry*, 1970, *11*, 49–62.

Sampson, E. E. The study of ordinal position: Antecedents and outcomes. In B. Maher (Ed.), *Progress in experimental personality research*, Vol. 2. New York: Academic Press, 1965.

Scarr-Salapatek, S. Race, social class, and I.Q. *Science*, 1971, *174*, 1285–1295.

Schaie, K. W. A reinterpretation of age related changes in cognitive structure and functioning. In L. R. Goulet & P. R. Baltes (Eds.), *Life-span developmental psychology*. New York: Academic Press, 1970.

Schaie, K. W., & Strother, C. R. A cross-sequential study of age changes in cognitive behavior. *Psychological Bulletin*, 1968, *70*, 671–680.

Schmidt, F. J., & Hunter, J. E. Racial and ethnic bias in psychological tests. *American Psychologist*, 1974, *29*, 1–8.

Scott, W. D., & Clothier, R. C. *Personnel Management*. Chicago: Shaw, 1923.

Scottish Council for Research in Education. *The trend of Scottish intelligence*. London: University of London Press, 1949.

Semler, I. J., & Iscoe, I. Structure of intelligence in Negro and white children. *Journal of Educational Psychology*, 1966, *57*, 326–336.

Sharp, S. E. Individual psychology: A study in psychological method. *American Journal of Psychology*, 1898–1899, *10*, 329–391.

Shields, J. *Monozygotic twins brought up apart and together*. London: Oxford University Press, 1962.

Shockley, W. Negro I.Q. deficit: Failure of a 'qualicious coincidence' model warrants new research proposals. *Review of Educational Research*, 1971, *41*, 227–248. (a)

Shockley, W. Hardy-Weinberg law generalized to estimate hybrid variance for Negro populations and reduce racial aspects of the environment-heredity uncertainty. *Proceedings of the National Academy of Sciences*, 1971, *68*, 1390A. (b)

Shockley, W. Models, mathematics, and the moral obligation to diagnose the origin of Negro I.Q. deficits. *Review of Educational Research*, 1971, *41*, 369–377. (c)

Shockley, W. Dysgenius, geneticity, raceology: A challenge to the intellectual responsibility of educators. *Phi Delta Kappa*, 1972, *53*, 297–307.

Shuey, A. M. *The testing of Negro intelligence*. New York: Social Sciences Press, 1966.

Singer, R. N. Interrelationship of physical, perceptual-motor and academic achievement variables in elementary school children. *Perceptual and Motor Skills*, 1968, *27*, 1323–1332.

Skodak, M., & Skeeks, H. M. A final follow-up study of 100 adopted children. *The Pedagogical Seminary and Journal of Genetic Psychology*, 1949, *75*, 85–125.

Smith, R. T. A comparison of socioenvironmental factors in monozygotic and dyzygotic twins, testing an assumption. In S. G. Vandenberg (Ed.), *Methods and goals in human behavior genetics*. New York: Academic Press, 1965.

Spearman, C. The proof and measurement of association between two things. *American Journal of Psychology*, 1904, *15*, 72–101. (a)

Spearman, C. General intelligence, objectively determined and measured. *American Journal of Psychology*, 1904, *15*, 201–293. (b)

Spearman, C. *The abilities of man*. New York: Macmillan, 1927.

Starch, D. The use and limitations of psychological tests. *Harvard Business Review,* 1922, *1,* 71–80.

Stein, Z., & Kassab, H. Nutrition. In J. Wortis (Ed.), *Mental retardation,* Vol. 2. New York: Grune and Stratton, 1970.

Stein, Z., Susser, M., Saenger, G., & Marolla, F. *Famine and human development.* New York: Oxford, 1975.

Stoch, M. B., & Smythe, P. M. Undernutrition during infancy and subsequent brain growth and intellectual development. In N. S. Scrimshaw and J. E. Gordon (Eds.), *Malnutrition learning and behavior.* Cambridge: MIT Press, 1968.

Stott, L. H., & Ball, R. S. Infant and preschool mental tests: Review and evaluation. *Monographs of the Society for Research in Child Development,* 1965, *30,* No. 101.

Tanner, J. M. Relation of body size, intelligence test scores, and social circumstances. In P. H. Mussen, J. Langer, & M. Covington (Eds.), *Trends and issues in developmental psychology.* New York: Holt, Rinehart, & Winston, 1969.

Tenopyr, M. L., Guilford, J. P., & Hoepfner, R. A factor analysis of symbolic memory abilities. Reports from the Psychological Laboratory, University of Southern California, No. 38, 1966.

Terman, L. M. (Ed.). *Genetic studies of genius. Volume I. Mental and physical traits of a thousand gifted children.* Stanford, California: Stanford University Press, 1925.

Terman, L. M., & Oden, M. H. *The gifted group at mid-life.* Stanford, California: Stanford University Press, 1959.

Thorndike, R. L. Review of *Pygmalion in the Classroom. American Educational Research Journal,* 1968, *5,* 708–711.

Thorndike, R. L. Concepts of cultural fairness. *Journal of Educational Measurement,* 1971, *8,* 63–70.

Thurstone, L. L. Multiple factor analysis. *Psychological Review,* 1931, *38,* 406–427.

Thurstone, L. L. *Primary mental abilities.* Chicago: The University of Chicago Press, 1938.

Tignor, B. W. *The relationship between minor physical anomalies and measures of attention and school achievement in primary school children.* Unpublished Ph.D. dissertation, Graduate Faculty, New School for Social Research, 1974.

Torrance, E. P. *Guiding creative talent.* Englewood Cliffs, New Jersey: Prentice-Hall, 1967.

Tuddenham, R. D. The nature and measurement of intelligence. In L. Postman (Ed.), *Psychology in the making.* New York: Knopf, 1962.

Tyler, L. *The psychology of human differences.* New York: Appleton-Century-Crofts, 1956.

Uzgiris, I. C., & Hunt, J. McV. An instrument for assessing infant psychological development. Mimeographed paper, Psychological Development Laboratory, University of Illinois, Urbana, 1966.

Vandenberg, S. G. A comparison of heritability estimates of U.S. Negro and white high school students. *Acta Geneticae Medicae et Gemellologiae,* 1970, *19,* 280–284.

Vernon, P. E. *The structure of human abilities* (2nd ed.). London: Methuen, 1961.

Viteles, M. S. Selecting cashiers and predicting length of service. *Journal of Personnel Research*, 1924, *2*, 467–473.

Waldrop, M. F., & Halverson, C. F. Minor physical anomalies and hyperactive behavior in young children. In J. Hellmuth (Ed.), *Exceptional infant: Studies in abnormality*. New York: Brunner-Mazel, 1971.

Wallach, M. A., & Wing, C. W. Jr. *The talented student*. New York: Holt, Rinehart, & Winston, 1969.

Wang, H. S. Cerebral correlates of intellectual function in senescence. In L. F. Jarvik, C. Eisdorfer, & J. E. Blum (Eds.), *Intellectual functioning in adults*. New York: Springer, 1973.

Wechsler, D. *The measurement of adult intelligence* (3rd ed.). Baltimore: Williams & Wilkins, 1944.

Wechsler, D. *The measurement and appraisal of adult intelligence*. Baltimore: Williams & Wilkins, 1958.

Weikart, D. P. *Preschool intervention: A preliminary report of the Perry Preschool Project*. Ann Arbor, Michigan: Campus Publishers, 1967.

Weiner, G. Scholastic achievement at age 12-13 of prematurely born infants. *The Journal of Special Education*, 1968, *2*, 237–249.

Werner, E. E., Bierman, J. M., & French, F. E. *The children of Kauai: A longitudinal study from the prenatal period to age ten*. Honolulu: University of Hawaii Press, 1971.

Whimbey, A. *Intelligence can be taught*. New York: Dutton, 1975.

Wiggins, J. *Personality and prediction: Principles of personality assessment*. Reading, Massachusetts: Addison-Wesley, 1973.

Wilkie, F. L., & Eisdorfer, C. Systemic disease and behavioral correlates. In L. F. Jarvik, C. Eisdorfer, & J. E. Blum (Eds.), *Intellectual functioning in adults*. New York: Springer, 1973.

Willerman, L., Naylor, A. F., & Myrianthopoulos, N. C. Intellectual development of children from interracial matings: Performance in infancy and at 4 years. *Behavior Genetics*, 1974, *4*, 83–90.

Wilson, P. T. A study of twins with special reference to heredity as a factor in determining differences in environment. *Human Biology*, 1934, *6*, 324–354.

Wilson, R. S. Testing infant intelligence. *Science*, 1973, *18*, 734–736.

Wilson, R. S. Twins: Patterns of cognitive development as measured on the Wechsler Preschool and Primary Scale of intelligence. *Developmental Psychology*, 1975, *11*, 126–134.

Winick, M., Brasel, J., & Rosso, P. Nutrition and cell growth. In M. Winick (Ed.), *Nutrition and development*. New York: Wiley, 1972.

Wissler, C. The correlation of mental and physical tests. *Psychological Review*, Monograph Supplement, 1901, *3*, No. 6.

Witty, P. A., & Jenkins, M. D. Intrarace testing and Negro intelligence. *Journal of Psychology*, 1936, *1*, 179–192.

Zajonc, R. B., & Markus, G. B. Birth order and intellectual development. *Psychological Review*, 1975, *82*, 74–88.

Zigler, E., Abelson, W. D., & Seitz, V. Motivational factors in the performance of economically disadvantaged children on the Peabody Picture Vocabulary Test. *Child Development*, 1973, *44*, 295–303.

Zimmerman, I. L., & Woo-Sam, J. M. *Clinical interpretation of the Wechsler Adult Intelligence Scale*. New York: Grune & Stratton, 1973.

Subject Index

EDUCATIONAL PSYCHOLOGY

continued from page ii

António Simões (ed.). The Bilingual Child: Research and Analysis of Existing Educational Themes

Gilbert R. Austin. Early Childhood Education: An International Perspective

Vernon L. Allen (ed.). Children as Teachers: Theory and Research on Tutoring

Joel R. Levin and Vernon L. Allen (eds.). Cognitive Learning in Children: Theories and Strategies

Donald E. P. Smith and others. A Technology of Reading and Writing (in four volumes).

Vol. 1. Learning to Read and Write: A Task Analysis (by Donald E. P. Smith)

Vol. 2. Criterion-Referenced Tests for Reading and Writing (by Judith M. Smith, Donald E. P. Smith, and James R. Brink)

Vol. 3. The Adaptive Classroom (by Donald E. P. Smith)

Vol. 4. Designing Instructional Tasks (by Judith M. Smith)

Phillip S. Strain, Thomas P. Cooke, and Tony Apolloni. Teaching Exceptional Children: Assessing and Modifying Social Behavior